INTELLIGENCE, RANDOMNESS,

AND THE STALEMATE

RUDY LEYERZAPF

ISBN: 1-4140-3870-4 (electronic)
ISBN: 1-4140-3869-0 (softcover)

Library of Congress Control Number: 2004090436

This book is printed on acid free paper.

Printed in the United States of America
Bloomington, In

1stBooks – rev. 03/10/04

To my wife, Ellen - my best friend for forty-three years

Contents

Preface

In the first month of the second year of the twentieth century, the Victorian age came to an official close. The era of Great Britain's longest reigning monarch had established strict – although mostly unwritten – rules on moral and "correct" conduct. One's "survival" in the world depended on adherence to the attitudes of respectability and appropriateness. Formality was upheld as a great virtue and anything below that, or less than that, was frowned upon – or worse.

Oh, how times have changed!

If Oprah Winfrey – needing no introduction to the reader in this part of the world – who openly used the once considered scandalous expression "p.o.-ed – the abbreviation of the words 'pretty offended (oh well, close enough)'" in one of her programs, would have publicly said that twenty years earlier, there wouldn't be a Harpo Productions Company today. And if the conservative President, who overtly blurted out the words "kick ass," would have been the leader of the free world in the nineteen-twenties instead of the nineteen-nineties, calls for his impeachment would have been a certainty.

Although there is never any excuse for mean-spirited vulgarism, it is now generally accepted that occasional "salty language" and "highly informal expressions" have their place here and there beyond the boundaries of private conversations. They do not only have the power to amuse the released-from-the-Victorian-straight-jacket individual, they also frequently "register" better than their formal counterparts. If the average citizen were asked a day later about the contents of the dispensed-on-the-picture-tube Presidential wisdom, the more honest one would probably answer: "He said 'kick ass.'" So, if you want your words to be remembered …, then every once in a while …. Well, like it or not, that's apparently how human brains are "wired"; and – sorry, well-bred and virtuous one – neither you nor I "hooked up that wiring."

The great genius of comedy, Robin Williams, whose delightfully extravagant comic outbursts are so enjoyable to watch, is one of my most favorite entertainers. The Kleenex box is usually empty at the end of one of his performances. But "my-oh-my," the words that roll out of his mouth from time to time would have made the great nineteenth-century queen – repeatedly mumbling "we are not amused" – get up and leave the theater immediately.

Shakespeare's saying: "To Thine own Self be true," has been in the public domain for centuries, but how many readers and listeners have "heard" it at the level where it really counts? But there is a not-so-lofty equivalent of the bard's maxim in existence that also expresses that same great truth of Man quite well. I have it in front of me in my office. The words are written in fancy scroll and mounted in a wooden frame. I think the anonymous author deserves a special-category Pulitzer Prize for them.

Here it goes:

> *Be what you is*
> *Cuz if you be what you ain't*
> *Then you ain't what you is*

Immortal wisdom, cast in – depending on your "distance" to the above-mentioned era – either amusing or unacceptable language.

Somehow, and in some strange way, that saying has helped me digest and appropriate that truth because it has "impacted" me more than Shakespeare's august and famous words.

Does that mean that I am fond of bad grammar and foul language? Most certainly not. A spicy expression may "fly" in isolated incidences when the point that is being made seems to benefit from it, but in general – particularly when it is overdone – it is a definite "no-Go" with me. However, there is such a thing as: "There is a time for everything under the sun …."

Here you have my excuse – motivated by a desire to reinforce an important point – for occasionally switching to the informal mode in the ensuing chapters. Don't worry, you won't find anything outrageous in them; unlike Robin Williams, I'm not *that* "courageous."

However, I have mustered enough courage to decide to leave some of the longer sentences in tact. Yes, my spellchecker has reminded me that I have used more than a few. I recognize the rationale for the shorter sentence. In fictional writing there is almost never a reason to use the longer ones. However, there is an undeniable disadvantage in separating – primarily for the loss of its effectiveness – an immediate important digression from a clause that it relates to by too many words. It is like tearing apart a complex mathematical formula and listing the components separately, or in series, instead of presenting them as an integral part of a complete unit. It's not quite the same. If there would have been any possibility at all to make the material of the book available to the Winnie-the-Pooh-level reader (this is not intended to be an insult), I – sincerely not wishing to shut out anyone – would have made a supreme effort to accommodate him or her; but, regrettably, that cannot be done. However, I believe that I have largely succeeded in keeping typical philosophical jargon – sometimes referred to as fancy gobbledygook – out of the text.

The reader should bear in mind that he or she has the distinct advantage over a listener at a lecture – when no tape programs are being sold – in being able to go back to the presented concepts over and over again. A complexity – the undeniable (but not to be feared) reality of life – should be seen as a series of more or less simple (or simpler) interconnecting ideas (like studying a road map where you look for only one street at a time instead of for all of them all at once) which can be assimilated over time by the intelligent man or woman with an appetite for it.

Introduction

This book deals with the age-old question: "Do we live in a Universe which came into existence as a result of randomly acting, non-intelligent forces, or is there an *a priori* Intelligence behind it all; and will it ever be possible to discover some substantial, or – ideally – irrefutable evidence for either model of Ultimate Reality"?

Throughout history, but primarily since the Darwinian Age, much literature has been written and many debates have been held attempting to answer and settle that – endlessly lingering – question. Several different approaches were taken: many of a religious nature – often laced with strings of warnings of dire consequences, others on purely philosophical grounds, and some on a presumed scientific – more than once unmasked as pseudo-scientific – basis. In most instances, arguments were made from a belief-system irreconcilably antagonistic to the one of the opponent. Adversarial and often intensely hostile forces armed with verbal weapons, usually fashioned from logically unsupportable leaps of faith, were marshaled to try to demolish the camp of the infidel, who, in time, would simply respond in kind. And in the end, most of the participants, as "unimpressed" and as "unconverted" as ever, would – after the verbal battle – still be members of the same philosophical clan. Aside from the benefit of sharpening one's rhetorical skills, lasting gains in understanding are almost never achieved by such an exchange between "us" and "them." Or, to put it bluntly and in a nutshell: from the perspective of making real and measurable progress in moving closer to the true answer of this most essential of all human questions, "another – interesting perhaps – but, nevertheless, futile exercise and fine waste of time on a ceaseless merry-go-round."

One would have expected that by now, well over two centuries after the publication of Immanuel Kant's *Critique of Pure Reason,* it would be universally understood (because the results are so completely predictable) that human reasoning – unaided by well-established scientific evidence – about the existence of an *a priori* Intelligence will always "come up short." But the much-beloved purely argumentative, searching-for-the-ultimate-checkmate method has never completely been abandoned in these post-Kantian times. In every instance, the method led to a stalemate and not to a checkmate; although some of the players may have labored under the illusion that they got to the latter instead of the former.

An example of a "high level (in the sense of well-argumented) debate" that fits into this category took place between the eminent British philosopher-mathematician, Lord Bertrand Russell (1872-1970), who was an agnostic, and Father F. C. Copleston SJ (a Jesuit priest) in the fall of 1948. The debate, which was broadcasted by the BBC, was also published in Bertrand Russell's book *Why I am not a Christian* (page 133). This exchange of ideas between these two brilliant men was, in essence, very much like a chess game. A good game – and great "stuff" for discussion and analysis, perhaps – in which some of the "maneuvering" may well be considered one of the all-time great "moves," but where "nothing of any consequence (from the standpoint of lasting philosophical progress)" really happened. When aluminum and magnesium are mixed in certain proportions, the alloy is stronger than either metal by itself. But meaningful discovery and additional strength in human insight and knowledge was, in my opinion, not a product of this exchange between these two learned men. Apart from the unmistakable shining eloquence by both parties, it was fundamentally all "old hat" that was being rehashed for the umpteenth time – and I seriously doubt that either man expected "to win the other one over."

Maybe it is time to "give up" – because it surely looks like "we're not getting anywhere" – on the failed, purely argumentative method which for so long has tried to "hunt down" the elusive answer to the above question. However, since that question isn't going away – ever – until satisfactorily answered, aren't we better off if we look squarely – preferably with the "mind-bias adjustment knob" set at neutral – at the absurd (for example in chapter 2: the identical twin phenomenon), but inescapable consequences of the extensions of the proposition that the totality of Man's consciousness is nothing more than a product of a "running" brain; and only then decide what is tenable and what isn't? Such an approach is unquestionably more effective. Hence the Why of this book!

Chapter One

The question of Ultimate Reality and Evolution

There are only three conceivable models of origin of Ultimate Reality. They can be described as follows:

1) There is an *a priori* Intelligence in or behind the Universe, which co-exists with Substance (particles and energy, in a myriad of forms and conditions), or from which Substance has emanated (it will most likely forever be impossible to know whether the former or the latter is true).

2) The Universe is filled with Substance *only* and there is absolutely nothing (no initiating or creative Intelligence) besides that.

3) There is no *a priori* Intelligence and there is no Substance – nothing other than pure nothingness.

Unlike possible models 1 and 2, model 3 is clearly nothing more than an imagined reality model that obviously never existed because the manifested Universe could not have come into being from that model. It is, nevertheless, an important "might-have-been" model for inclusion in our considerations.

1

Both possible models, model 1 and model 2, include the incredibly-difficult-to-fathom concept of beginninglessness, or of being uncaused. The idea that something lasts forever and ever into the future can, with a little reflection, be processed and accepted by the brain. However, an attempt to fully grasp the concept of an endlessly-moving-further-and-further-back-into-the-past-without-ever-coming-across-a-starting-point-anywhere-along-the-way reality – which is what beginninglessness means – can almost short-circuit the human brain, because it is totally conditioned to think in terms of cause and effect relationships only. The Substance in model 2 is, rightly viewed, an effect without a preceding cause – and, of course, so is the *a priori* Intelligence in model 1.

The Big Bang – an expression jokingly coined by somebody who didn't particularly care for the hypothesis – which may well have been the starting-point of the manifested Universe (after several decades the theory still seems to be holding firm according to the majority opinion of astronomers and astrophysicists), was certainly not the moment of creation of this primordial Substance – but rather a point of conversion. Of course, it remains to be seen if we will ever know anything meaningful about the reality of the Substance before what is commonly referred to as the Birth of the Universe and the first appearance of Space-Time and Matter as we now know it. The opinion – and of course that's all it is – has been expressed that the Big Bang may not have been an explosion at all, as the name suggests, but rather a point when super-hot Substance began to flow and expand in all directions from the size of a pin. What is a virtual certainty, however, is that we cannot really say that this mysterious, most-likely-still-existing-today (there is no reason to believe that all of it was "converted" – or that it is even depletable) pre-Big-Bang Substance is an integral part of our familiar three-dimensional reality, because it somehow formed (or caused) that (Space-Time) reality – and the effect and its cause are never one and the same thing. The ordinary cause and effect relationship principle may not – in view of the possibility of the existence of such realities as anti-matter, etc. – have existed prior to the Big Bang, but I think it's safe to say that it already pertained to the just-formed three dimensional reality at the moment of – and certainly after – the Big Bang.

However, because there isn't a whole lot more that one can presently say about this pre-Big-Bang existence of this pre-Matter Substance, this subject matter is, understandably, given very little attention outside limited scientific circles. Perhaps, the "String theory" with its accompanying theory of "Supersymmetry," developed by some theoretical physicists – which, according to insiders, looks promising mathematically and has now gained a substantial following – along with additional knowledge of black holes, will eventually shed more light on the pre-Big-Bang reality. It is hoped that the giant particle accelerator or cyclotron, which is now being built in Switzerland (we started building one right here in the United States but the project was stopped due to lack of funding) will eventually verify those theories.

Fascinating as this all may be, it, nonetheless, remains an indisputable fact that if there were no Substance in whatever form, knowable or unknowable, moments as well as forever and ever – or perhaps more accurately stated "timelessly" – before the Big Bang, there simply wouldn't have been a Big Bang. Model 3 couldn't have produced a Big Bang, because it is an indisputable axiom that "Absolute Nothingness can *never* become or produce Something." Only Something – again, in whatever form or condition – can become, or change into Something else. So, all this "Stuff" that we are looking at, and with which we are looking, hasn't just been around for a mere 15 billion years (in February of 2003, it was re-calculated to 13.7 billion years), or whatever the age of the physical Universe happens to be, it has been around in one "recycled" form or another eternally. Unlike what Einstein initially believed, the manifested Universe probably had a definite beginning, but this ultimate, irreducible Substance *never did*. I suppose most of us rarely think of it that way. So, nothing was ever "started," but everything is forever "changing" – and now you can see why that is not merely a cliché.

Whoever would claim not to have at least a bit of an intellectual problem with the idea of "being uncaused," has never spent much time contemplating it. Accepting it over time, or a gradual getting used to the idea is, of course, completely different from comprehending it; but it can be sensed intuitively. The frequently expressed skeptical reaction upon confrontation with this idea of beginninglessness: "But who then created this *a priori* Intelligence?" – or if asked from a religious perspective: "Who then created God?" – points to the difficulty and discomfort in digesting that concept. However, if it is rejected on the basis of it being – as one student in a philosophy class unceremoniously put it – "too wacky," it needs to be borne in mind that its only possible alternative, model 2, may well be considered slightly "wackier" in that respect. For is it really less taxing on our limited human reasoning faculties to accept the picture of beginningless Substance (including [post-Big-Bang?] "fermions" and "bosons," "quarks" coupled to "s-quarks ["s" for sister particles in a parallel dimension according to the Supersymmetry theory]," and whatever as-yet-undiscovered, possibly more subtle, perhaps non-three-dimensional particles or energy forms may be added in the future) forever and ever "popping around" in an otherwise "non-vital" Cosmos – which, according to the model 2 view, for no apparent reason just happens to be there, but, without a creative agency, was never ever "put" there – rather than the idea of an eternal *a priori* Intelligence and First Cause existing outside (or independent from) time and space? Yes, either model is an enormous stretch to the limited contemplating human mind, but the model 2 concept unquestionably requires – "let's get real here" – an *impossible* stretching. Ironically, only the obviously incorrect model 3 is the "easy" and "logical" one, for it would have been far more (humanly) logical if nothing ever existed, but that is obviously not the case. But no matter "how we slice it," at the ultimate point of origin, where things can be reduced no further, there is a Reality which was never "made," and is thus uncaused.

True, for various reasons one may have a greater affinity with model 2 than model 1, however, nobody can point to an inherent intellectual necessity – because beginninglessness is as much an aspect of model 2 as it is of model 1 – to accept model 2 and reject model 1 based on this concept of beginninglessness. It (the necessity) simply doesn't exist.

Perhaps the ultimate question

A question of stupendous importance lies before us: "Will it ever be possible to prove the correctness of model 1, and in doing so eliminate model 2, or vice versa, or will success in this direction continue to elude Mankind"?

Up until now it appears that, like the gold-miner, the philosopher could do no better than to continue to sift through the sand and pebbles of mostly pure speculation in search of that solid nugget which would elevate his writings on the subject to the level of true knowledge and independently verifiable truth. Unfortunately, it appears that all these "head on" attempts to prove model 1 have so far remained largely unfruitful. After a respectable endeavor was presented to the world, another thinker would come along to point out the shortcomings and weaknesses in the argument and expose the lack of "watertightness" in its reasoning. Certainly, the perceptive critic, such as David Hume in his well-known critique of St. Anselm's ontological argument, deserves our applause for showing us the flaws in a perhaps otherwise beautiful treatise.

So, why has it been so difficult up until now? Is there something in the Cosmos that prevents us from finding the answers to these basic questions in life? Not according to Einstein, who said, "Nature is not vindictive…, it's just that She is so far ahead of us."

What about if we avoid the "head on" approach altogether and do not waste our energies on trying to find the one single perfect – but terribly elusive and possibly even non-existent – argument that is supposed to wipe out for all time all the objections and counter-arguments in one fell swoop. Why not look in various directions instead to see if an accumulation of pertinent material, cautiously weighed and conservatively evaluated, will not give us, if not a 100% ironclad proof, then at least a strong indication in which direction and how far the scales are tipping? The miner hoping to see a gold nugget in his pan, but who only collects gold dust over time, still has his payday.

The weighing process

The beginning position – before we have placed anything on either side – of the figurative (weigh) scales is, of course, perfect balance (the needle is in the center), because there is nothing in the description of either model that adds weight to its own validity.

Beginning with model 2, can any qualifying solid evidence in support of this model be added to its side, which is *not* – and I repeat *not* – a mere part of a belief-system?

Model 2 is, in a nutshell, the statement of belief of the atheist. It's probably fair to say that most people – as well as many atheists themselves – view atheism as a frame of mind of non-believing. Nothing could be further from the truth however. For how can anyone claim that he or she doesn't believe in anything when there remains any kind of believing at all? Only the true agnostic with his in-the-final-analysis-it's-not-knowable disposition is the genuine non-believer, but a real atheist (or non-theist) is an active disbeliever – and there is quite a difference – and therefore entertains a formulated belief-system. Surely, it is a belief-system of denial, but a belief-system nevertheless, which – perhaps much to the chagrin of the proponents of it – inherently doesn't belong to the realm of true knowledge as has so often been suggested. Why? Because for atheism to be elevated out of

6

the realm of believing (which is nothing other than merely suspecting things to be a certain way), with its typical logically unsupportable leaps of faith, into the domain of knowledge and certainty, would require a full-blown – and nothing less than that – disproof or refutation of model 1. That, dear reader, would be infinitely harder to do than to disprove – or to prove for that matter – that there is intelligent life on a planet, if it even exists out there, in a galaxy a billion light years away from earth. To prove or disprove such a reality is completely beyond the present-day capabilities of astrophysicists.

Beliefs and the processional caterpillar

If I may take the liberty to digress a little bit about the reality of a belief (creed) or a complete belief-system, a valid comparison can be made with the findings of the anthropoid research done by the French entomologist, Jean Henri Casimir Fabre (1823-1915), who – according to his detailed studies outlined in his book *The Life of the caterpillar* – discovered that if he lined up processional caterpillars head-to-tail in a circle around a flowerpot, with plenty of food nearby, they – with their powerful follower instinct – would continue to march around that pot until they eventually collapsed from exhaustion and starvation. Many beliefs, including the atheist's, are closed-loop thought patterns similar to the endlessly repetitious marching regimen of the processional caterpillar; and – coming to think of it – isn't that essentially a fine vicious circle? The preacher would say: "The Bible is the only revealed Word of God." One courage soul may have the audacity to ask: "But how do we know that?" His answer: "Because the Bible says so." That's a superb example of a closed-loop thought pattern, which for the time being seems to satisfy the non-researching, easily-put-off, never-question-religious-authority type of human being, but it doesn't do much for the determined man or woman who really wants to find out – demanding diamond hard evidence all the way along – "what's out there." A belief, without any form of proof, may be more or less comforting

7

to the believer – and that fact undeniably constitutes its temporary value – but that value, nevertheless, pales in comparison to the permanent significance of thoroughly tested knowledge. Ideally, the latter should eventually completely replace the former (and there is little doubt that it gradually will), if for no other reason than that there are many widely varying beliefs – religious and otherwise – with more than a few in arrogant and ugly packages, but there is only one uniting Truth with its many extensions – which, more likely than not, won't resemble any particular doctrinal belief-system; and only true relevant science, correctly interpreted, can take Mankind to that point, where absolutely nothing else can. We know, for example, how an electric motor works and fully understand the underlying natural laws relevant to its working; and that understanding is the same for every mechanically and electrically aware person on the planet. So, we may speak – in a completely non-mystical way – about the "uniting truth or knowledge of the electric motor," where if we, collectively, didn't have a clue of how it works, a wide variety of opinions – most of them patently false – would exist; although, unlike the repugnant-consequences-of-the-divisions-of-creeds phenomenon, the differences in opinions (if it were possible that they could continue to exist) about electric motors wouldn't matter all that much.

True knowledge unites because it pushes all disparate opinions into oblivion; and that is why science has to be married to philosophy and religion, so it can add some sorely needed sanity to the business of the latter two!

Ever been invited to a "2+2=4 Revival"?!

A medical doctor's opinion of garlic and onions

An example of an unexamined closed-loop belief involves a medical doctor's unfortunate opinion of the nutritional value of garlic and onions. Because I am not interested in mounting a personal attack against anyone in this writing, he will remain anonymous. On one of his tape programs, he stated, categorically, that garlic and onions should be avoided in one's diet because the bulbs grow below the soil's surface where sunrays play no direct part in the growth of these vegetables. He said that his conclusions were based on ancient Hindu scriptures.

I am not a nutritionist. However, I have read enough books and articles on the subject of nutrition to know that there is a general consensus among the various experts in that field which totally contradicts the rooted-in-India's-antiquated-wisdom doctor's opinion. Now, I am a westerner who had the privilege of visiting India two decades ago and who truly appreciates India's wisdom in general. But, apparently, the ancient Hindu writer(s) – if that is what he, she, or they actually wrote – didn't "get it right every time," because reliable research has shown repeatedly that garlic – and onions to a lesser degree – has, among other benign aspects, valuable blood pressure regulating properties, is not known to be toxic in any way, and, as a natural food, has – barring the "social odor problem" – no side effects. That fact should remind us that it's *never* a good idea to take any so-called wisdom – ancient or modern – on face value. The Oracle of Delphi – imagining for a moment that it really did have the ability to speak – may have spoken humbug instead of Truth. And aren't we to think of Truth – and not of Socrates!? The Thing Itself is supposed to stand up on its own … apart from who says what!

Shouldn't the individual, who presents himself (or herself) as a person of insight to the general public, make sure that no misinformation is ever dispensed. If information is controversial and untested – then: say so! To me, that is a sacred duty you have to your audience and readers. Why be eager to pick up a banner that is suspect, which may make you – embarrassingly – march in the wrong parade?

A polemic agnostic

Bertrand Russell, the well-known British philosopher-mathematician, wrote in *Why I am not a Christian*: "I am an agnostic and not an atheist because in the final analysis it would be impossible to disprove the existence of God." (paraphrased;

page 133 and elsewhere)

Again, unlike the more neutral agnostic, an atheist is someone who – without a shred of scientific evidence supporting his views – actively embraces Reality model 2. So, do the discoveries that support the theory of evolution not qualify as proof of the correctness of model 2 – giving the atheist his modest tangible "shred" to hold on to? I wouldn't mind being generous and magnanimous here, but the facts still dictate the answer: "Most certainly and emphatically not." Certain interpretations of what the findings mean may suggest that the phenomenon of evolution is only consistent with model 2, but the problem here is with the interpretations and not with the findings themselves. Figuratively speaking, the latter are in black and white and the former are colorized. To counter these interpretations, many defendants of their (mostly religious) version of model 1 have made an all-out attempt to try to reason the theory of evolution into oblivion, hoping that in doing so their model as the remaining survivor would come out on top. If, however, one is not hampered by the snares of a belief-system – whether religious or irreligious – then an unbiased observation of what is "scientifically out there" becomes possible, which should lead to the realization that it isn't at all necessary to deny the fossil records, because the records themselves do not in any way prescribe or decree the direction in which they must be interpreted; only the mindset of the interpreter will do that. Natural selection is observable and doesn't need to be denied, but whether or not it is nothing more than a product of randomly working forces in nature, as opposed to possibly being one of the – perhaps several – modi operandi of an all pervasive Intelligence behind it, the phenomenon itself cannot tell us.

DNA and the mutation

In the twentieth century, Mankind was introduced to the "modifying factor (or cause)" in evolution: the mutation. But why did the non-disease-causing mutation ever have to be labeled a chance mutation? Do we really know how chancy it actually is?

Modern science has revealed how extraordinarily complex the DNA directed building process is. As far as we now know (from the recently completed Human Genome Project), the three billion (yes, with a "b") letter human DNA code controls the production process of a variety of incredibly complex, "three-dimensionally interconnecting and interacting" proteins, which appear to be the basic parts or building blocks of what we are biologically.

When a mutation occurs, there is a spontaneous (unknown causes) sequential change in the letter code of the DNA, creating a rearrangement in the protein structure. But how does DNA "know" how to change its own code so that a useful biological modification becoming an evolutionary advancement comes out at the other end?

Certainly, one must acknowledge here that as yet unexplainable, relatively rare breakdowns becoming disease-causing mutations (for example, a few letters "disappearing" from the DNA sequence) do occur, but that fact doesn't in any way deny or downgrade the astounding phenomenon of the "successful" mutation.

While I am discussing these DNA issues, I was reminded by a TV news commentary that it was exactly fifty years ago that the coil-shaped structure and the copying process of the DNA molecule was discovered by a 24-year-old American, James Watson – who made the unflattering remark about "the entrenched scientist": "A goodly number of scientists are not only narrow-minded and dull but also just stupid" (Ouch!) – and a 35-year-old Englishman, Francis Crick. Ever since that time, expressions were (and still are) used in reference to that discovery, such as: "The secret of Life now unlocked …, etc."

Well, certainly, the scientist unlocked *one* secret, which gave us access to highly beneficial applications of that knowledge – and I share in the excitement of that important discovery – but, haven't we simply pushed back the questions a little further in the directions of *all the other* secrets behind that *one* (former) secret? Shouldn't the words: "… a part of …" be inserted in the right spot in that exuberant expression? To suggest that we now know it all – which may not have been the intention behind the expression – is blatantly arrogant in addition to being woefully ignorant (the respectable equivalent of Watson's word "stupid").

So, could it (the above-mentioned appearance of the successful mutation) all summarily be dismissed as being nothing more than "just another run of the law of probability" – even though this level of complexity reminds us that there is, to say the least, a highly significant deviation of what we could reasonably have expected this randomness law to produce? Statisticians tell us that every once in a while deviations in the law of probability do show up. However, to insist that millions of them – and there have been that many mutations in the course of evolution – with their combined odds running incalculably high, can still be made to fit into this law surpasses all levels of believability. A man or a woman of intelligence and intellectual integrity wouldn't accept these odds as being possible in most other areas in life. So why accept them here? Only those individuals totally trapped in their allegiance to a belief-system would claim that such overwhelming odds against their favorite model don't bother them in the least.

The gradual change in evolution

But one may ask: "Doesn't the idea of a gradual change over a long period of time, like the snail-like erosion forces that shaped the Grand Canyon for example, justify the placement of the process of natural selection under model 2?

Well, there is indeed a gradual element in the process of natural selection, namely the passing on of the more successful mutated gene(s) down from the first "better equipped" individual to a larger group, which is then "fitter" and therefore superior to its cousins without the mutation and consequently better able to survive. Of course, the expansion through breeding from a single mutated individual is a slow – at least at the outset – process. However, we can also identify a sudden event or abrupt change in natural selection, because the first spontaneous appearance of the mutation must take place between a *single* set of parents and their immediate offspring; and there is nothing gradual about that! Why? Because the non-mutated parent must – in a "single leap" – produce offspring with a complete, fully functional mutation. I trust that the intelligent reader can, without much further elaboration, readily see that a new sequence in a successful survival-of-the-fittest process can obviously not start on a "half-baked" or a "still-in-the-developmental-stages" mutation. It is also now widely believed that many of the evolutionary adaptations (resulting from mutations) were a shorter series of major modifications – or bigger leaps – rather than a longer series of minor ones.

So, again, it is this suddenly-emerging-within-a-single-generation, highly complex mutation, coupled with its abundant – although all different and unique – manifestations in evolution that should give us pause to reflect on the reasonableness of its (the gradual change consideration) placement under either model. It's safe to say here that the odds heavily favor model 1!

The disappearance of the dinosaur

An example of an unnecessary and completely unwarranted spin added to a dramatic event in the history of evolution involves the reporting of the discovery of an enormous impact into the earth's surface by a very large meteor or asteroid off the coast of the Yucatan Peninsula, Mexico, some 60 plus million years ago, which presumably caused the disappearance of the dinosaur from our planet.

This interesting and educational story has been presented on some TV programs in recent years, however, regrettably, with the usual commentary in the narration: "reminding the viewers how close ALL of life came to being completely wiped out as a result of the climatic change following this 'catastrophic' event." But was it really such "a dumb random event," or "a doomsday near miss"? Well, yes, if we are shortsightedly stuck in bemoaning the fate of the dinosaur!

However, if, with our eyes wide-open, we begin to see that this event may have provided the very opportunity for other forms of life to flourish, which might not have been possible with the continued presence of these incredibly large beasts roaming the prehistoric landscape, then this ancient occurrence doesn't have to be quite so shocking as the sensationalized commentary suggests (of course, we now have to live with the "Steven Spielberg ones," but the only real estate they take up is a little bit of virtual space in the human imagination and a few Gigabytes on a DVD disk; and those video species represent no threat to Man or beast except in a possible nightmare).

Regardless of whether evolution is "meant (model 1) to produce" or "just happens (model 2) to produce" more advanced forms of life – such as beings that can go to concerts, or compose great symphonies for them, etc. – this so-called catastrophe should properly be viewed as an incident triggering a desirable – intimating the presence of intelligence far more than the absence of it – change in direction which evolution needed to take. Catastrophes rarely produce primarily constructive results – and if they ever do, it is perfectly asinine to identify them by that word (or name them so).

Now, we certainly have no *definitive* way of establishing – the avoided-throughout-this-book introduction of a "belief" being impermissible here – if it was "meant to produce," or "just happened to produce" subsequent evolutionary advancements. However, we simply now do know that things turned out well for the mammal class of animals and their further evolution all the way till the present day following the demise of the dinosaur. That's a true triumph story for "Life" (applicable to both models) again and again overcoming and succeeding, isn't it?

So, in light of the fact that evolution is still such a hotly debated and contentious issue, particularly around some of our elementary educational institutions, is there really any constructive purpose served in this slanted doomsday talk other than a possible dubious benefit to the presenters of such programs for satisfying the appetite for drama and sensation of a segment of the viewing public?

Every day in America and in most of the western world, millions of cars are passing each other from opposite directions at a very high combined speed. But that fact, fortunately, rarely leads to frontal collisions because of the presence of intelligence and control in the drivers, who, with the exception of a few mentally unstable individuals, do not go home reporting to their families that they came close to dying that day simply because they passed an opposite-direction large truck on the road within a few feet distance.

No, we don't know (non-belief knowing) and shouldn't claim to know – and I cannot stress that enough – if there was any presence of "higher" Intelligence and control around that event 60 plus million years ago – but – neither do we know that there was *not*; and therefore cannot eliminate that as a possibility with absolute certainty. To do so would be strictly unscientific. That is one of the so many things that we, collectively, do not know and should acknowledge as such. It must remain standing on the shelf until … – not because I demand it, but because the facts themselves demand it – and whoever removes that without further substantial eliminating (model 1) evidence is simply not worthy of the name "scientist."

At this point, I would like to state a self-evident fact that theoretically doesn't even need to be discussed because in principle it is so blatantly obvious, and yet so many otherwise intelligent people – even though they give intellectual assent to that fact – seem to overlook, or ignore it in their arguments from time to time. Here it is: If an *a priori* Intelligence *was* present all the way along (detectable or not) – at the time of the Big Bang and at the time of the asteroid impact (and, of course, then at any other time) – all the collective states of mind of denial of all the atheists and agnostics on the planet *cannot remove* such an Intelligence from the

Universe – and conversely – if It never existed, all the collective affirmative states of mind of all the "believers" in the world *cannot put it in*. Human beings do not have that kind of power.

So, if we are incapable of proving the affirmative of a possibility, does that mean that – what so many logically undisciplined people seem to think they must do – we are compelled to accept its alternative or the negative? Of course not! Surely, the thoughtful man or woman with a deep-seated integrity will realize that the process of "enlightenment in these matters" is never aided by ignorant – although perhaps unintentional – spins in any direction.

Other than at the favorite game and in a few other areas in life, Mankind doesn't need to be "spoon-fed" informed about who's winning or losing during the delivery of such public programs. But it does have a genuine need for unbiased presentations of important discoveries which can only come from a die-hard healthy attitude of suspending judgment when and where we don't really know. Why not adopt the fair-minded and common sense policy of leaving it all "right smack in the middle"? It certainly would invite less criticism and unnecessary controversy.

A spin in a different direction

A few years ago, I read an article by an all-fired-up creationist trying to convince his readers that the theory of evolution had been rendered obsolete because human remains were found right next to dinosaur bones. I don't know if he made the whole thing up, but let's assume for a moment that his story was accurate. Is that discovery a reason to abandon the widely accepted concept of evolution and – probably what the author of the article was secretly hoping for – adopt creationism and wear its badge and uniform? I don't think so. There may be perfectly reasonable explanations why skeletons and fossils of creatures that lived millions of years apart are found side by side. Here is a possible one: a dinosaur dies and his

16

skeleton is preserved in the geological stratum belonging to that period. Millions of years later, erosion forces "bring the skeleton back to the surface (many dinosaur bones have been found near the surface or were even partially exposed)." Then a primitive man goes hunting in that same area – but this time his powerful quarry is victorious, and the man dies. A possible scenario …? There you have it: fossils and remains nearby. However, if we find the teeth marks of a Tyrannosaurus Rex in a human skull – then it's obviously time to reconsider some time frames. And even then there is still no need to scrap the *whole* theory of evolution. However, if it were to come to that – and I consider that highly unlikely – my morning Espresso would still taste "every bit as good."

I'd rather quietly read what Nature is trying to tell us – than zealously concoct a reality theory based on grossly misinterpreted evidence.

Letting Randomness "muddle" with the creative process of building a fin on a pre-fish

Somewhere along the line in the evolutionary process, the transition from an earlier life-form to a fish was completed. Because I am not an evolutionary biologist, I will only use crude images to describe the stages of that transition.

We know that the amoebae – and the same is true of many later life-forms – did not have much, if any, control over the direction in which it moved, if it moved at all, through its environment. If it moved – or perhaps more correctly: was moved – it did so as a result of outside forces, such as water currents, etc. Then there comes a time when evolution "wants" to modify or advance the form into a fish. We'll let Randomness or Pure Chance wrestle with that project to see what it can do.

Left-handed Randomness, with its – using the amusing eloquence of our youth: "great-big-fat-zero" IQ, begins by accidentally putting a strange growth on the outside of the body of the immobile and rudderless creature. Well, it's not exactly a handsome – more like resembling the drawing of an airplane of a three-year old – streamlined paddling device, but, "what the heck," what more could we realistically expect from our "designer"? Although it appeared in a poor location on the body, it'll have to do for now. So, let's "start it up" to see if it paddles. Oops, there are no muscles! Now we must wait, and wait, and wait some more, to see if Randomness can, again accidentally – of course many generations down the road – attach a muscle to our paddling device. Aaaaaahhh, but the paddle had "fallen off" in the mean time, because the paddle mutation wasn't "favored" by a survival-of-the-fittest process which obviously couldn't have been activated by our "non-advantageous," non-functional paddle. The rule in natural selection is that whatever hasn't reached the stage of "being favored" is doomed to disappear.

But, being die-hard model 2 optimists, we keep rooting for the appearance of another – this time "ready to go" with a muscle attached to it – paddle. Random-ness "racks its brain," oh sorry, "non-existent brain," and low and behold against all pathetically high odds, a primitive paddle with a muscle appears. Well, let's start that one up to see if we fare any better this time. "Oh shoot," there is no nerve connection to give the "Go" signal. However, that too "comes in" against now truly staggering, unpronounceable, and insanely wild odds.

And then, of course, brainless Randomness still has to go through the formi-dable task of building the return-to-the-forward-position-with-a-minimal-amount-of-drag-through-the-water-controlled-by-brain-programming-to-ensure-correct-movement muscles to get that – one – paddle or fin working.

Yes, one fin only! Why did we so carelessly think that two symmetrically placed fins – so the creature isn't doomed to forever go around in circles – would have been built in the *best* location on the body by our "devoid of any intelligence" Randomness? A gross oversight perhaps? A slothful habit of thought maybe? Does the idea of grandiose absurdity apply here anywhere?

Can anyone come up with a good explanation why these one-fin-in-the-wrong-location creatures and billions of other similar trial and error "screw-ups" have never been found in the fossil record? They certainly should have been there if it all happened accidentally and chaotically! In addition to asking ourselves questions about what we do find, it also seems like a good idea to me to be equally curious about all the things we do *not* find. The latter reality often appears to be – conveniently? – overlooked by the average atheistic evolutionary biologist.

Yes, that is the way in which Randomness would have done it, because that is the best that "It" can do. Unlike the antique car restoration enthusiast who enjoys the luxury of working a few hours here and there on his two-year-project in his garage, Randomness had to feverishly manufacture and assemble its adaptation *overnight* in order to take its extraordinary complex project to the necessary-for-survival "favored" status, otherwise the whole "shebang (its project)" – unlike the not-subject-to-natural-selection-principles, not-yet-assembled car – would have simply disappeared in a short time. It was an all or nothing proposition many millions of years ago.

So, if there is – using the process of elimination – a 99.999999…% certainty that some type of intelligence must have been involved in the "fin building," why would any well-thinking, non-indoctrinated, without-a-blindspot individual even for a moment entertain the notion that model 2 might be the correct model of Reality? "Uh, hmmm, duh, …!"

It is because of the details that the reasonableness of the pure chance atheistic choice disappears. Agnosticism has more credibility than atheism!

Chapter Two

The question of Core Awareness

If model 2 (Chapter 1) were the correct model of the Universe, then Man's total consciousness cannot be anything more than a mere product of brain functions. That's an inescapable conclusion. In that case, there is obviously nothing that can survive the disintegration of the biological entity. That state of "being," or more accurately "non-being," is like reality model 3 (Chapter 1), where there is no awareness or experience whatsoever, or, in other words, a total nothingness; not even an awareness that there is no awareness.

The late Carl Sagan, scientist, author of best-selling books, such as *Cosmos* and *the Dragons of Eden*, as well as former TV host of programs relating to science subjects, wrote in an article in "Scientific American" magazine that he was convinced that human consciousness is only a product, or resultant of the brain, consistent with what is written in the preceding sentences. However, in that same article, he said that, like most human beings, he wished that he could believe that his parents were still alive in another dimension along with the possibility of future reunification, but added that he considered that prospect highly unlikely. Although a greatly respected and popular scientist, it is clear that what he wrote in that article was an expression of his personal beliefs and nothing that comes close to the classification of well-established or proven human knowledge.

Like the whole "raison d'être" of this book, the challenge is to see if we can find relevant evidence that will take us far beyond the ethereal nature of all beliefs, including Carl Sagan's, and, if not all the way, then at least a good distance in the direction of that just-mentioned classification.

The essential part of living beings

The most amazing part of our consciousness is by far our basic awareness, meaning: our capacity to experience. That a brain, which developed over eons of time, can reason is one thing, but that there is also the experience of the reasoning is something else. Artificial intelligence, capable of beating world champion chess players and carrying on an intelligent conversation with human beings, has been around for some time now. However, unlike us humans, it's utterly incapable of experiencing or enjoying either the chess game or the conversation. It can "do" – but remains totally unaware of its "doing"!

So, what do we mean by basic or core awareness? The following scenario may be helpful in explaining, and giving us a feel for what it is and what it is not.

A normal and intelligent man (or woman) sits on his patio with his dog beside him and his six-months-old son or daughter in a baby chair. While they are sitting there, the neighbor's cat walks into the yard, and all three beings watch the animal. What are some of the thoughts and emotions, or "mind movements" that may occur, which ones will occur, and which ones cannot occur in each one of the three beings?

a) The normal man will, in addition to his awareness of the presence of the cat, recognize both the type of animal and the fact that it is the neighbor's cat (unless, of course, he has never seen it before). He may possibly remember its name, may remember that this cat caught a mouse a week earlier, and also may feel a certain amount of resentment over the fact that the neighbor does little to prevent the cat from walking into his yard.

b) Because of the obviously aroused interest in his six-months-old baby, we can see that it is every bit as aware of the presence of the cat as the father is. However, because of its early stage of brain development, it cannot recognize or know the type of animal, let alone the cat's individual history or name. Apart from the baby's curiosity and corresponding emotion, it has no further "mind movements," such as a memory association.

c) His dog is also keenly aware of the presence of the cat, which clearly shows by the wagging of his tail. It is obvious that he is looking at an animal that he would love to chase, however, without the superior mentality of his owner, he is incapable of attaching a name to the cat, and placing the animal in the feline class of mammals.

(In other writings, the words consciousness and awareness are frequently used interchangeably. However, for our purposes, the word consciousness will be used in a wider sense, meaning: awareness plus the various brain functions.)

As can be clearly observed, in all three living beings (humans and animal) there is an awareness, and, of course, an awareness of something. This awareness is unique and exclusive, because no living being can have a direct experience with someone else's awareness. We can only climb into our own observation tower but we are never given the key to somebody else's tower!

In case of the father, there is reasoning and there are memory associations which are completely absent in the infant. As an intelligent man, the father, unlike his dog and infant, can also reflect on his own existence, and thus possesses the capacity for self-awareness. So, there is a more expanded consciousness in the parent compared to that in the child. However, we cannot say that the father had a greater direct awareness of the presence of the cat than either the dog or the infant. That demonstrates that memory and other brain functions, including even self-awareness, have little, if anything, to do with core awareness. In fact, the all-important question here – worthy of further scrutiny – is whether or not this core awareness is a brain function at all.

To illustrate this further, another "observer" should be added to our trio on the patio. Because this observer is so totally different from the other three, "it" joins them at a later moment. This observer is the most sophisticated, state-of-the-art robot that scientists and artificial intelligence engineers have put together in recent years. It is equipped with "eyes" similar to camera lenses which the robot can focus on whatever it "chooses," or is programmed to look at. The robot "decides" to look at the cat. It subsequently "consults" its extensive database in search of matching data that will allow it to interpret what it is looking at. A split second later, its "speaker voice" says: "I'm looking at a domestic cat, it is so much smaller than its distant cousin, the tiger."

This impressive product of human engineering was clearly able to do something that two of the other observers, the baby and the dog, could not do. It "knew" what it was looking at. However, unlike the baby and the dog, it did not "experience" its seeing and its knowing. There is "nobody home" in such a machine.

Although, many brain researchers no longer say that the human brain works exactly like a computer, it is still believed – and probably rightly so – that brain functions are every bit as mechanistic as what is going on inside the "head" of the robot. However, the thoughtful scientist realizes, or should realize, that not even the most brilliant of men have any clue *whatsoever* as to how a brain activity translates into an awareness. How does the "somebody that *is* home" – the core

23

awareness in the three living observers – make an experience out of a brain wave? That is still – to say the least – a highly perplexing mystery. Only those researchers who suffer a little bit too much from the proverbial "contempt bred by familiarity" would put that on the shelf as an insignificant problem soon to be explained by mechanistic means.

It is that elusive something, which is present in living beings and conspicuously absent in the robot, that we can identify as basic or core awareness.

A helpful analogy would be to compare human consciousness to a Broadway production. During the performance of a show, such as a musical, we have two distinct groups of people inside the theater: the active actors and singers on the stage and the passive audience, or spectators in their seats. If we liken brain activity or functions to what's happening on the stage, then core awareness may be likened to the watching audience. In the normal human experience the show is complete because the performance (the brain activity) is delivered to the audience (the core awareness). However, inside the robot there is *only* a non-audience-attended rehearsal going on, because the stage activity – as marvelous and elaborate as that may be – is *not* being "delivered." Nobody is watching!

Yes, it is unquestionably correct to say that we have learned a lot about the brain in recent years. However, it seems to me that all the learning so far has only to do with what is happening on the stage and little, if anything, with the audience. The main reason for that is that activity – on the stage or in the brain – can be monitored; but how does one monitor the "non-active (or inactive) watching" of the still-sitting audience or the "non-active watching" of core awareness? If the audience is watching – as they are expected to do – the performance in the "physically inactive mode," why is it unreasonable to think that pure core awareness may be watching the experiences of life in the "mentally inactive mode" (brain waves for everything except for core awareness itself)? No matter how hard anyone may

24

try, that possibility cannot be excluded.

That possibilities cannot always be excluded also applies to other "phenomena of the mind." I personally suspect that ghost sightings are "in the mind of the observer" sightings (real enough to him or her) rather than "retina registered" sightings, but I don't think we have any definitive way of excluding the possibility of the latter.

Core awareness and cloning

Another example may shed some more light on the question of core awareness. We live in a time when there is much talk about human cloning with all the ethical ramifications associated with that concept.

Our objective is not to enter into a debate on this most controversial subject. However, for our purposes it is useful to reflect on the reality of "growing human bodies and minds from genetically identical cells." We do not have to wait until a humanly engineered clone will make its first appearance – which, rightly or wrongly (high probability of severe abnormalities), is bound to happen sooner or later (according to the latest news, it is already on its way) – before we can learn something about this phenomenon, because Nature's human clone, the identical twin, has been around since the dawn of human history.

In our example we will take a closer look at John, David, and Steven, who are identical triplets. A question arises concerning their core awareness: "Although everything that they are biologically was "built" from totally identical DNA, does that mean that their core awareness is also totally identical and that we can only account for their identities being distinctive, or one of a kind, because the core awareness "happens" in separate localities, namely their individual brains? There is absolutely no doubt that they are separate human entities because, although there is usually a strong bond between identical twins and triplets, John definitely knows himself to be a non-David as well as a non-Steven, David knows himself to be a non-John and a non-Steven, and Steven knows himself to be a non-John and a non-David.

That leads us to another interesting question: "What happens when John dies and then his preserved DNA were to be revived by scientific methods, which, although presently unknown or perhaps not yet entirely feasible, may become available to us in the near future? That is certainly not outlandish, or far-fetched, because the whole cryonics industry is based on that concept. So who would that new individual be? Is it John? Is it John back into the business of "experiencing" again because his core awareness is the same as it was before he died? Not necessarily so! It may not be John at all!

Identical twins or triplets are formed in the womb because of a relatively uncommon dividing action in the developing cells from a single fertilized egg, occurring at some point after conception. The further developing, but now separate, twins or triplets (or potentially more) have, nonetheless, identical and indistinguishable DNA in their genes; hence the name "identical twins." That means that developing cells with identical DNA not only *can* but definitely *will,* at times, produce different and separate individuals. Although John's DNA produced John himself of course, it is also correct to say that John's DNA produced David, since both of them are a biological product of the very same fertilized egg.

So, who will it be then that is being revived from John's DNA? David? Steven? Or John himself? And what about if it is David's core awareness "up and running again" and the original David is still alive? Do we then have one individual in two places at once and, if so, how would that be experienced? That's impossible to imagine, isn't it?

We are not being overly confident in stating that this simple example clearly shows that the core awareness in human beings is totally unique in itself and that its uniqueness is not just a product of the fact that the core awareness resides in its separate housing, namely the human brain.

These considerations inescapably represent an insurmountable problem when Man is looked at as just a living machine running on a genetic program arising from model 2 (chapter 1). If on the other hand this core awareness in Man is an independent unit (only possible under model 1, chapter 1) instead of a mere product of the functioning brain, then the possibility of a single individual living simultaneously inside two separate brains does not exist. But it clearly and irrefutably does exist under model 2!

The cryonics philosophy

The famous baseball player, Ted Williams, recently (2002) passed away. It was reported in the news that, apparently against his expressed will, his body, or a part of it – like Walt Disney's remains – was frozen in a cryonics laboratory for possible future revival.

But the same question "pops up" again: "Which Williams will it be who would be coming back"? Will it be Ted himself, or Bert Williams, his "might-have-been" twin brother, or Randy Williams, Bob Williams, Sam Williams, Jim Williams, Bill Williams and a host of other "theoretically possible" identical Williams brothers? Ted Williams may not have had an actual identical twin brother in life, but – as we all do – he certainly had "potential" identical siblings (there is no biological reason to think that we cannot all be cloned), and anyone of them could "occupy" the newly revived and running physiological system which was actually intended for Ted's return. There is no special marker on the DNA that reads: "To be used for the benefit of the original owner only and not for his identical twin."

And how many potential identical siblings do we all have anyway? Hundreds? Thousands? More? The human womb has physical limitations, but such limitations do not exist in a cloning laboratory. What a blow it would be to the cryonics industry, once more and more potential customers would begin to reflect more deeply on the reality of the foregoing. If, for instance, you – the reader – would consider having your remains, or part of them, frozen after your death, and the actual number of your potential identical siblings or clones would be, let's say, one thousand (highly unlikely under model 1, but quite possible under model 2), then your chances of being revived through the use of your DNA are only one in a thousand, because there are nine hundred and ninety-nine potential "non-yous (of course, the plural of you with an 's' doesn't exist; I only momentarily created it to make the point)," and – only one you. Not very good odds, wouldn't you say!

This very strange who-is-it-that-comes-back reality can be reinforced by another example involving identical twins or triplets.

The stereo example

In this instance, we will place two totally identical (same brands, same models, etc.) stereo systems a distance apart, for example, one in Chicago and the other one in Los Angeles (the actual distance is not relevant, however, they cannot be electronically linked together). We then place two (one for each city) completely identical CD's, for example one of Mozart's beautiful piano concertos, into the trays of the CD-players, and start playing them at a precisely synchronized moment. It is obvious then that Mozart's music played in Chicago is also heard in Chicago and, of course, the same holds true for Los Angeles. However, what is also unquestionably true, not as a matter of speech but as a matter of fact, is that the CD played in Chicago is also heard – note for note – in Los Angeles. If the "hardware (stereo)" is the same and the "software (CD)" is the same, we know – because there is nothing mysterious going on here – that the resultant product (music) must also be the same.

To make the comparison between the stereo phenomenon and the identical twin phenomenon, we will only use John and David this time. If we consider the brain reality for the identical twins under model 2 (chapter 1), we realize that either everything (the DNA as well as the total brain configuration [whatever it is that produces this "resultant" core awareness]) is identical – or – the brain configuration has become unique for each of the two brothers, in which case we are at a total loss explaining what might have caused the differentiation, or diversification between the time of the cell separation and the fully grown brain of each of the twins; and, yes, even more so with the further diversification in case of triplets, etc. Natural selection obviously couldn't have caused the differentiation, because the survival-of-the-fittest-by-passing-on-superior-genes process doesn't take place in just one single individual between the time of conception and reaching adulthood. There are also no other known natural forces or phenomena, capable of overriding the standard DNA instructions for exact cell duplication in the identical siblings, which might have created a unique diversification in the brain in the short period of time between conception and full brain maturity. That makes the notion of brain configuration diversification in the twins, or triplets, an exceedingly unlikely candidate to explain the awareness uniqueness of these otherwise totally identical human beings.

So, if we place John, with his identical-to-David DNA and brain configuration (like the stereo equipment), in the room with the stereo (or in any other room for that matter) in Chicago, and David, with his identical-to-John etc., in the room in L.A., and both of them are awake and alert (like the CD's being played), will it then be correct to say that John, while experiencing with his own core awareness (like hearing the music played locally), is also experiencing with David's core awareness (like hearing the music played in the distant city), or, in other words, is inside David's head as well as inside his own and vice versa. Well, yes, it must be correct, because, as with the stereos, no mysterious or esoterical element is present in this – consistent with model 2 – "purely mechanistic" running of the brain's "hardware and software" and its resultant "totally-non-*a-priori*" experienced awareness. No matter what brain researchers may ever come up with, as long as John's "gray mass" is, atom for atom or even quark for quark, identical with what's inside David's skull, the example will hold true – because everything is and remains identical (now the brain configuration diversification theory in the twins has been shown to be untenable) from beginning to end. Totally identical causes, or series of causes, can only produce totally identical effects and there is no way around it! Now, at no time in all of recorded history has any identical twin that ever lived reported having had the being-in-two-places-at-once experience! The foregoing seems to suggest that we may well have stumbled on the outright impossibility and absurdity of the extensions of model 2.

This second example of the "twin phenomenon" leads us to the same result as the first one; and thus the reinforcement.

If it is too confusing

These identical twin examples may be somewhat confusing to the reader. Fortunately, there is another direct way to illustrate this very odd being-in-two-places-at-once experience. Again, it is a model 2 (chapter 1) reality *only* of course, and – I cannot stress this enough – it doesn't apply in any way to model 1. It is useful to cast it in the form of a brief "mind exercise."

I (generic/nobody in particular) was born in the nineteen-twenties and I die in the year 2000. A substantial amount of my DNA is preserved. A scientist using my DNA "brings me back" in the year 2020 (if it produced "Me" once, it makes no sense to deny that it can do it again!). The DNA that produced me in the nineteen-twenties, produces me again a hundred years later (we'll ignore the identical twin possibility in this instance; besides, since there is a lot of my DNA, the scientist could produce a few identical siblings before he gets to another "Me" anyway, so the reappearance of "Me" is a virtual certainty). So, because of the successful DNA manipulation, I am growing up again in the 2020's, just as I did in the 1920's. Aaaahh, but there is so much more of my DNA. With the billions of remaining cells, there is nothing to prevent the scientist from repeating the process and starting up another "Me." Then I am literally born again in the year 2021. Which one of the two then is the real "Me," the one born in 2020 or the one born in 2021? BOTH of "us" are/is "Me." Wow! Again, the core awareness has been exactly duplicated. There are now definitely two "Me's" running around in separate bodies – BUT – we all intuitively know that – THAT CANNOT BE! Again, this example indirectly tells us that model 2 (chapter1) cannot be the correct model of the Universe. However, when a non-duplicable "independent experiencer or core awareness (model 1, chapter 1) – whatever 'stuff' it may be made of" – joins the developing biological entity, then the second biological entity from the same DNA is joined by "another independent experiencer (a non-Me)," or a model 1 identical twin.

A second exercise

Let's do another exercise to "drive the point home" even further. This one is also a model 2 *only* exercise. It has the added advantage of demonstrating to the interested "student of the mind" that consciousness is far more "real" than physicality. This exercise must be done in the "first person mode" or the "I am mode."

I ("Me," the author, or "You," the reader, etc.) sit in my chair and contemplate my own existence. I look at the sunshine and realize that the sunlight is there because I am aware of it (philosophical "Idealism") and determine in my reasoning that the sunlight is also there because it is reasonable to think that the sunrays have independent-from-my-own-awareness existence (philosophical "Realism"). I die (don't worry, you're not going to "keel over" because there is no suggestive power here!) tonight and my awareness "goes out," and because of that fact there is no more sunshine – only from the perspective of philosophical Idealism – tomorrow morning (of course there is also no more "Idealistic" tomorrow morning!). But does the sunshine – or more accurately my sunshine observation – then stay on, or suddenly come back, because I have an identical twin brother who is still alive? Of course not. My "I-awareness" is not *his* "O(to me)-awareness ("O" for Other).

(When we are both alive, I do not run on the beach with him while I am sitting at home reading the newspaper. I can imagine that I am with him, but that is not the same as a "direct experience"; and as I have said earlier in the chapter, such a "direct experience" has never been reported by any of the millions of identical twins).

Now, the sunshine also doesn't come back to me in my deceased state because of my identical triplet brother's awareness; because he also has an "O- to me but I- to him" awareness.

So, now we have one very distinct "I-awareness (mine)" and two "O-aware-ness (brothers) [there is no plural form of the word: "awareness"!]."

Most of us would accept the fact that every other identical sibling or clone of mine – possibly millions or more – "produced" while I am alive, must have an "O-awareness." We do not think that Dolly the sheep in the barn is "with" the "cloned from" mother in the field. Those two may be identical in every other way but not in core awareness. But if that is true, it would be virtually impossible for "Me" or "I-awareness" – because of the virtually unlimited number of potential O's – to be "revived" in my "belonging-to-me DNA package." But isn't that exceedingly strange, that those odds were not against me when I was born? I readily "popped into" that DNA package then. Doesn't that also show the "weirdness" of the model 2 approach in a decisive way?

Doing such an exercise is like building a sandcastle on the beach at low tide when you know very well that the incoming tide is going to sweep it all away. When the basic premises you build from are false, the structure must cave in. Nothing that you try to "set up" under model 2 makes any sense whatsoever!

Could there possibly be anything else that "writes"?

On December 22, 2001, a cloned cat was born. Apparently the cat looks quite different from its "parent"; according to the reports, there is also little behavioral resemblance. Now how in the world could that be? Without a father there are no new genes; and DNA, like a photocopy machine, is only supposed to be able to copy in exact duplication; and we cannot ascribe to it any ability to alter the image or to "re-write the script." If "mistakes – meaning faulty duplications" are made in the copying process, that should lead to imperfections, breakdowns and diseases – and *not* to wholesome modifications. There is no way, unless one continues to indulge in intellectual fraudulence (that's not the kind for which you can go to jail!) or self-deception, that one can maintain – as the absolute reductionist has staunchly done until now – that DNA is the "sole supreme writer and producer of the – animal or human – individual." There is clearly something of an elusive nature that writes along with the DNA that does have the power of modification. If DNA alone was in charge, the cat would have been "the spittin' image" of its SINGLE parent.

So, to sum it up, either the core awareness in Man is nothing other than an outgrowth of the functioning brain, in which case it is at least potentially possible for one individual to experience himself in two localities simultaneously, in other words, the core awareness is then duplicable – or – the core awareness has an independent side to it and is essentially immaterial or "non-three-dimensional" in nature – whatever that means – in which case, of course, core awareness duplication simply doesn't happen.

Even though we have to admit that we have no way of establishing what this core awareness is in essence, nobody can successfully argue that the former makes more sense than the latter. It's most definitely the other way around!

Riley's ghost in the machine

I have little doubt that my fellow (psychology and philosophy) bookworms have come across the one-time raging debate over "Riley's ghost in the machine." Oh, how the ultimate reductionist tried to clobber the poor ghost into pulp! But wasn't that a little bit like fighting an elusive pitch-black creature in total darkness? How can one even go after it, when there is no knowledge of what it might even be?

Could it have been core awareness all the way along? I'll leave that an open question.

Strong recent scientific evidence for a metaphysical reality

In the first chapter, I have briefly alluded to the widely supported "String theory" of the Universe. Within that theory – or more accurately "theories," because there are several – the theory of Supersymmetry was developed and formulated by theoretical physicists. The latter theory strongly suggests that for every three-dimensional particle there is a companion, or sister particle in a higher-numbered (fourth, fifth, sixth or whatever) dimension that somehow interacts with its three-dimensional counterpart (there is a large Internet website – easily found by typing the words "string theory" into a search engine – that informs the layman as well the physicist on this subject).

At any rate, one necessary extension of the sister particle concept suggests that there is an exact duplicate dimension – not far away, but existing alongside our familiar three-dimensional world (although naturally invisible) – in which everything that exists in this world has a carbon copy reality in another universe or dimension. The Hindus, among others, have believed in that concept for thousands of years, but now the core of that belief is receiving some serious scientific attention and support. So, to accept the possibility of a metaphysical (beyond the physical realm) reality is no longer as "unscientifically crazy" as it was once made out to be – quite the opposite.

Now the die-hard skeptic who coined the phrase: "I don't believe in anything of which I cannot build a scale model," will – if he is honest – have to eat some three-dimensional crow here; and I'm sure that expression suggests that the black bird isn't very tasty. The better informed individual might answer with: "Why don't you build your scale model out of dark matter – which is an unseen non-three-dimensional substance known (by its effects) to astrophysicists – and have your son or daughter take it to school for "show and tell." Surely, it's good to be a skeptic – and I frequently rank myself as one – but the brighter ones among us will place a cap on that skepticism. It must have a limit.

So, in all likelihood – it may take some time before this mathematically-sound theory is experimentally verified – there is a duplicate "you (and this is not your identical twin)" walking alongside your physical self (like your shadow on a sunny day) in a fourth, fifth, sixth, seventh, or whatever dimension (according to the "string" calculations there are ten dimensions including our known one).

Could it be that Core awareness – which, in spite of exhaustive research, has *never* been "located" in the brain – primarily "resides" in one of these higher di-mensions? Could it be that your non-three-dimensional reality (or self) is – like the cause and effect relationship between the non-three-dimensional pre-Big-Bang Substance and the three-dimensional post-Big-Bang Universe – more causative to your three-dimensional reality than the other way around? And could it be that these non-three-dimensional realities of human beings (ourselves included, of course) are more permanent than the three-dimensional ones?

I cannot presently answer these questions with a definite "Yes." I'm merely raising the questions. But the informed skeptic can no longer answer them with a convincing "No" (the uninformed ones will probably keep on doing what they've always done); we are well past that point now.

With the presence of this "mathematically handsome" Supersymmetry theory, no one – apart from the occasional outright intellectual idiot – can ridicule these legitimate, not-yet-answered questions back into the realm of fables, mysteries and the occult. Yes, these are harsh words, but they are justified here!

I'll put on a party hat and bring a toast on the day when they fire up the particle accelerator I referred to in chapter one.

Summary question:

Have we now not established that Core Awareness, or "just the Thing that watches" – which has nothing to do with, or has no essential dependence on memories, reasoning, a sense of self-identity (dogs and cats probably don't have that – but they do have core awareness), the functioning of the eye (the robot has eyes – but there is no core awareness), or any other biological function, all of which have a connection to brain functions – must be a non-resultant, or *a priori* immaterial unit (and I don't care what particular name we attach to it)?

Deny that – and you have an incredibly bizarre situation on your hands!

P.S. A certain amount of redundancy along with a fair measure of elaboration was intentionally used in this chapter, because, although the examples are relatively simple, the absorption and the digestion of the meaning of them may not be a quick process at all. Moreover, because this subject material typically invites some in-depth prodding and probing to test the structural soundness of the synthesis, which is welcomed, a high degree of thoroughness became necessary to increase the likelihood of this synthesis passing the crucial test of legitimacy and accuracy in the mind of the discerning reader.

Chapter Three

The question of Failures in Evolution

In chapter one, Lord Bertrand Russell, the British philosopher-mathematician, was quoted as saying that he considered himself an agnostic and not an atheist. He stated in his book *Why I am not a Christian,* that one of the reasons that he was inclined to doubt the existence of a Supreme Being was that he couldn't reconcile the phenomenon of a Ku-Klux-Klan member, a Fascist, or an Adolf Hitler, etc., with a purposeful and intelligent creative process (in the preface [page 9] and again on page 18).

It is not difficult to see why these members of the human race – not infrequently regarded as sub-human – who clearly are the antithesis of what the psychologist, Abraham Maslow, called "a fully self-actualized human being," represented such a philosophical enigma to Lord Russell.

I trust that the reader will agree that everything that has been discussed so far in the first two chapters has only added some weight to the side of the scales of model 1 (chapter 1). Is this the first time, because a totally impartial evaluation process seems to demand it, that we must add some weight to the side belonging to model 2, because the "Hitler phenomenon" has created – for us, as it had for Bertrand Russell – an outright philosophical impasse? A great hurdle, yes, but not necessarily an impasse!

In order to be able to jump over that hurdle successfully, it is necessary to relate in some detail to the image of a Deity that the eminent philosopher most likely had in mind while he wrote down his statement of rejection. This Deity or Creator, according to that image, fashions his creatures similar to the way a manufacturer designs and fabricates a complex product. If this manufacturer turns out defective or shoddy merchandise on a regular basis, the buying public is correct in accusing him of incompetence. So, Lord Russell must have reasoned, perfectly logically, that since Hitler was an obvious defective product of the Creator – which demonstrated to him the incompetence of the Deity – the idea of Its/His existence is therefore an absurdity, because incompetence in a Supreme Intelligence and Power is a strict contradiction.

So far, consistent with our perspective, theological speculations have been (and will be) scrupulously avoided. Will it be necessary now to abandon this perspective temporarily in order to show that alternatives to Bertrand Russell's objected-to-image are possible? Not really!

If we go back and continue to study the evolutionary process carefully, a clearly observable picture of a stage beyond the physical level of evolution, or the evolution of form, begins to emerge. That stage may be called the evolution of consciousness.

Man's participation is required

The universal process of biological evolution has outfitted most human beings with a potentially highly capable brain. But the recipient had nothing whatsoever to do with the development of it. Similar to a modern computer, the brain – as the hardware – has some pre-programmed software installed on it. Perhaps, we should refer to that as Man's basic survival character, and He, from the standpoint of making conscious choices, had very little to do with that development either.

Now, we shouldn't be surprised – precisely because it was an evolutionary process that took Man from the savanna into modern times – that some raw emotions, impulses, and reflexes, etc., which had served him well in the past, but are no longer as necessary as they once were, have not yet been "deleted (or modified)" from the average brain's programming, because the process of "deleting these unwanted files" does involve conscious choices and effort in most instances; they rarely "fall off" by themselves.

Aggression is an example of one of these raw emotions. Is aggression inherently good or bad? It's hard to imagine how the Neanderthal could have survived and protected his family without it. The civilized option of calling 911 was not available to him, so he had to be handy with his club. And what about if the grizzly bear knew no aggression? Would the bears then be "happily and lovingly" bunched together, like ants in an anthill, instead of being spread out over vast territories?

So, it isn't true that aggression is bad per se (if it were, we would then have a serious problem reconciling that with model 1), however, what should be obvious is that modern Man needs, if any at all, a much smaller dosage of it than his cave-dwelling ancestor.

This process of reducing – and perhaps ultimately eliminating – aggression while putting a much-more-suitable-for-modern-living "mind quality" in its place, such as tolerance, is an example of the evolution (or improvement) of consciousness. Many excellent books have been written on how that may be accomplished.

It is essential, however, that we realize that, unlike during the involuntary biological evolutionary process, our full participation in that "second stage" of evolution is required and no measurable progress can be made without it. The forward thrust is no longer universal at this new stage and it must be transferred to, or picked up by the individual if it is to continue. Nature produced the hardware, but we *ourselves* must write, re-write, and improve the software; and because it is entirely an internal affair, nothing will be – or even can be – superimposed upon our writing by any outside agency. If we want to learn to drive a car, we must get behind the wheel at some point and drive while learning the driving, and no one – including deities, angels and what have you – can do the learning for us. It simply cannot be bypassed in any way!

As in biological evolution, where we witness the spreading of mutants from single cells (or groups) and isolated locations, we see a similar unevenness or disparity in the development of consciousness between members of the human race. It is not surprising then that some (or even many) of them have not yet done a lot of re-writing at their stage. Some folks are "getting it" earlier than others. Like towards the end of a marathon race when all the runners have spread out, some human beings will walk (or run) in front in the van of evolution, the majority will be found in the center, and some will be ambling in the rear. It is obvious what section Adolf Hitler and Joseph Stalin walked in during their lifetime.

So, the "failures" – horrible and inexcusable as they were – in these men, which earned them their placement in the rear echelon of human evolution, doesn't necessarily have to be interpreted as evidence of a "design flaw," or some kind of "Universal botching," because, perhaps, human beings are not a "finished" product out of some sort of cosmic quality-controlled manufacturing plant. The man or woman, who refuses – possibly on religious grounds – to spend time reflecting on this development process, will continue to walk around with enigmatic and perplexing beliefs and images that are completely unnecessary to entertain. However, once the awareness begins to dawn inside a human being that wisdom *can* be – with definite thought pattern restructuring methods learned and applied during Man's life-cycle, and not between the clouds or in a Walhalla – but – *never* could have been "dumped into him" as some type of instant liquid goodness poured into his skull – because that's not how evolution works – then these seemingly impenetrable beliefs will begin to disappear. It then begins to make sense to him or her – a major revelation to many, I'm sure – that all the way to Maslow's self-actualization point of the individual evolution of consciousness process, the mind of Man will be and even *must* be – yes, you're reading right: *must* be – lopsided (when some character qualities are well-developed while others are not; or are still entirely missing). It would have been dangerous with possible fatal consequences, for example, for primitive human males to have been as nurturing as their women because that would undoubtedly have interfered with their aggressive reflexes which were so necessary for the protection of the family unit and the species. The eventually-to-be-developed "balanced personality" was *not* needed – or even welcome at that early point. The unbalanced personality is *not* a mistake of Nature (the Intelligence behind it being infinitely smarter than we are) at that stage, but it does have to be *personally* transcended at some point after that stage. Evolution demands it. Progress demands it.

So, where is the room for all this awful "guilt stuff" that the indoctrinated human being is so preoccupied with? There are far better ways to encourage people to grow out of these not-in-any-way-created-by-the-individual-himself – largely subconscious – primitive personalities! Man is at the core of his being not guilty, because how can we rightly blame him for his deeply-rooted Nature-ordained lopsidedness? Only the shallow thinker would ask: "How come men are so ..., or women are so ...?" – but the deeper thinker understands these unavoidable temporary phenomena that are to be worked out in the experience of living.

But, it *is* true that post-savanna Man *must* change – antediluvian behavior not being acceptable in modern times – and is *required* to adapt to the "new conditions," even if, paradoxically, He, notwithstanding the rank beginner in the self-actualization process that He is, has to – in the absence of gentler, and by far preferable, well-working methods – undergo some type of appropriate (even harsh) punishment to do so. The little bit more sophisticated ones among us will recognize that there is no contradiction in principle here.

Armed with the clear logic of the foregoing, and without the need for any special "religious or mystical revelation," we can now draw the legitimate conclusion that Man during this life-cycle is still very much "under construction."

We must add learning to enjoying

So, "life" is – in addition to enjoying it – about learning many things. Most people spend an inordinate amount of time "recreating themselves" – and there is nothing whatsoever wrong with the idea of recreation – and, if any at all, very little time in "re-creating themselves (notice the hyphen)." But "Self re-creation" is the most essential business of life – while "recreation" is a legitimate "fringe benefit." The latter is important – but not preeminently so!

One of those things, which must be included in the learning process, is the art of "learning to live." That is a surprise to many because they – largely subconsciously, and thus without any contemplation – had assumed all the way along that, because they were doing it (being alive), they must know something about the business of living and be relatively good at it. However, when we take inventory of the quality of life of the average human being, it becomes obvious that such is not the case. The majority of men and women still walk around with the notion that happiness and success are locked in things outside of themselves and are supposed to be hunted down rather than – using a both-ancient-and-modern slogan – "to be created from within." I would like to spend a paragraph or two explaining – which is rarely, if ever, done by the teachers and preachers who are working hard trying to convince their audiences that the latter is "the way to go" – why that strong tendency exists in Man. I suspect that "the explanation" has potentially more power to effect a change than "the (unexplained) persuasion."

Is it true that the reliance on things and conditions outside of us must eventually be transferred to something inside of us? Absolutely – because the "more advanced lifestyle" requires it. We need to return to our life on the savanna for a moment to make sense of that last statement.

Unlike in the example below, where an artist overcomes diabetes, success – and its accompanying emotion of jubilation – for our prehistoric ancestor unquestionably existed for a large part outside of him. He was successful when he carried his freshly killed quarry on his shoulders back to the group where others were waiting to process the meat and hide. Yes – in the absence of being able to make a good living writing poetry – his "fortune" had to be stalked; and "it" had the annoying habit of trying to stay outside the range of the javelin, or bow and arrow. Undoubtedly, his failures – like most animal predator attacks – far outnumbered his successes (are those failure and hardship memories the inspiration for the myth of Sisyphus, the King of Corinth, who was condemned by Zeus to push a stone uphill that forever rolled back?).

No wonder – because Man lived on the savannas so much longer than between skyscrapers – the thought patterns of the "chase of the hunt (his prosperity dream)" and "the source of his subsistence being outside of him – far away in the hills somewhere" were burned deeply into the human psyche. Charles Dickens' fortune – financial and otherwise – didn't come from something he had to hunt after physically, and neither did Elvis Presley's. These two men (and all other successful ones like them) – whether they consciously realized it or not – had "to BE something (produce something from desires, imagery, capacities, tendencies, etc. within themselves)" rather than focus on trying "to GET something (or physically run after it)." That luxury was not available thousands of years ago – and may still not be for the Papua in New Guinea.

So, this having-to-hunt-success-down-rather-than-creating-it-within notion is none other than a perfectly logical evolutionary carry-over from the past, and for that reason – although sorely in need of modification and correction – nothing to be embarrassed, or feel guilty about. And – "listen up" – that's not the only carry-over. There is a whole string of them. The human tendencies of laziness, procrastination, and other forms of inaction are carry-over's from the days when the conservation of energy principle called for as much inactivity as possible – like the whole lion pride taking long naps after feeding – in order to make explosive bursts of energy available during the hunt.

And then there is the lack of initiative – an endlessly waiting for something, so common in so many people, which is a subconscious memory from the days when a prehistoric protocol dictated that all the clan members wait – however long it would take – for a clear go-ahead action signal from their chief.

And next, the more disturbing and unacceptable behavior of ancient and modern times, called stealing, can be directly linked to the decreed-by-survival-pressures logical urge to scavenge. Relax, understanding the origin of this anti-social behavior doesn't – as is so often assumed – lead to excusing or tolerating it. Far from it!

And here is another big – no – huge one: Man's difficult time with monogamy has an undeniable connection to the easily observable promiscuity of the non-guilty chimpanzee – our closest ancestor in the animal kingdom. That the wolf is by nature more monogamous than Man (a puzzling phenomenon), doesn't reflect favorably on any "virtuous" individual wolf, as the pious Tartuffe would have it – or – unfavorably on Man. Certainly – because of the consideration of a more ideal arrangement (family, trust in an intimate partnership, etc.) – He can learn, with contemplation and effort, to override that archaic survival-driven, not-personally-chosen promiscuous instinct; but that He will falter a time or two along the

44

way is a given.

And there are many more – too numerous to list here!

Now, it should be obvious to the uncontaminated mind that none of these yet-to-be-modified powerful evolutionary tendencies and vestigial behaviors are a product of a conscious, willful disobedience to a Divine mandate. It is incredibly ignorant, or one might even go so far as to say: totally insane, to blame the Man of the savannas – who had no choice but to follow his survival instincts, who also had little, if any, opportunity to exercise his newly acquired larger brain (with its rudimentary, and therefore weak decision making capacity) for fancy reflective purposes – for having made the "one-time apple seduction mistake" in a "Garden" in which he had never lived. Nice try, theologian! No wonder the brilliant Bertrand Russell – and many others like him – didn't "swallow any of that stuff."

Yes, agreed, he (generically) *has to change* – and that's what evolution is all about anyway – but not because he was ever "guilty at the core of his being." That wretched guilt business is tommyrot, a lot of bull, a bunch of crock, and a load of crap, sprinkled over with holy water! It is legitimate to label it as such, because that belief – coupled with the bloodthirsty and revengeful belief of what Man has to do to fix his guilt – mocks the Universe and makes Its Author, if there is one, look like a psychotic barbarian, and – in modern slang – "a very sick puppy"!! The labeling isn't obscene – the belief is! Every Kindergarten graduate knows that you cannot condemn a creature which had no input in its own creation, is compelled to act out impulses it doesn't (yet) understand, and is only slowly beginning to wake up to its own essence. (Remember, I'm going "after the jugular" of a crazy belief [while you're hugging it, you may not realize the true nature of it, but if you – possibly having fought your way through your conditioning – stand back a little, then …] – which, like these behaviors, is badly in need of an overhaul – and NOT after the [figurative] jugular of any believer. It's not personal in any way! To tear a belief

45

apart is legitimate – because if it was based on Truth, which is indestructible, it'll survive anyway – to tear a person apart *never* is!)

The words of "the Ghost of Christmas Past" of the *Christmas Carol* by the above-mentioned author, are applicable here: "that these shadows are the way they are, don't blame me …."

Neither the promiscuous chimpanzee – nor prehistoric Man – nor you – nor I – created the Universe. So, if you feel compelled to react, here is a convenient stratagem: blame "the Man upstairs" – if you are convinced there is one – and see "what kind of reaction you're gonna get."

Suffering from diabetes

Learning and enjoying are actually tied together, because an increased involvement in the "true learning process (there is such a thing as useless learning)" can actually increase the capacity to enjoy life. However, a prolonged ignoring of the need – or actually even stronger: the requirement to learn, expand, or evolve – can lead to disastrous consequences, as may be seen in the next example of a "failure."

A woman in her forties had been suffering from diabetes for over a decade. During that period of time, she had felt a potent urge to fulfill her life-long dream of being an artist with paint and brush. She also suffered from the all too common "handicap of the mind" of not acting on these inner prompters. The usual "excuses to one's self," such as lack of time and money, etc., got in the way, and there never seemed to be enough courage to jump across this gape of insecurity. However, one day she started – as she had done in earlier years – doodling on a piece of paper, and suddenly the created image looked "so neat" to her that she decided to apply paint to it and expand it into a work of art worthy of wall space in an art-gallery. With the momentum thus gained, she kept going, and at age fifty had become a well-established artist. Although this is a true-life story of an actual person, it is, of course, not an altogether uncommon one, because there have been many people who went through similar experiences. What is unusual – but not unparalleled – however, is that shortly after the production of that first painting her blood sugar and insulin level – without doing anything special such as changing one's diet, etc. – had returned to normal; and she has remained free from the disease ever since.

Now, what was the reality of the disease in her? This so-called "demon" left her rapidly once the transition to a closer-to-her-heart edition of herself had been made. The aspiration to a fuller enjoyment of life was clearly held in check by – none other – than her own inaction or "disobedience" to the "inner calling" of a far-less-mundane-than-what-she-had-known-up-until-then reality within. This strongly suggests that if she had taken care of her "inner" problem (overcoming her procrastination) earlier, the "outer" problem (the diabetes), with all the – to put it mildly – unpleasant attempts to fix it externally, might never have been there in the first place.

One may choose to call this rather "mystical" stuff, but that isn't necessary at all! It's plain to see that capturing a beautiful image on canvass, or engaging in any other creative endeavor is a far more enjoyable and emotionally rewarding activity than performing menial tasks, or doing any routine job that can be done well by a machine, such as digging ditches, etc. Surely, something within us "drives" us – but without overriding anything – to abandon the lesser for the greater. The Neanderthal, being too preoccupied with his immediate survival, wasn't "there yet," but many of us in modern times have plenty of opportunity to "answer" these inner prompters long before "something has to hit us over the head." So, if we can only dump this no longer needed pre-historic conservation of energy thing, then ….

A handicap that actually helped?

The next example of a so-called "failure" is also of a medical nature. Around the turn of the nineteenth century, the composer, Ludwig van Beethoven, began to develop hearing problems. Two decades later his doctor declared him totally deaf. His deafness was so complete that when he conducted his last symphony, the famous Ninth or "Choral" – which, in itself, is an astounding accomplishment considering the nature of that handicap – he was totally unaware of the immediate standing ovation and thunderous applause from the audience. One member of the orchestra gently took the composer by the arm to nudge him to turn around, so that he could see the overwhelming appreciation from the Viennese music-loving public for the composer himself, as well as for his magnificent, recently completed work.

So, wasn't that loss of hearing a clear-cut tragedy, particularly for such a great man of music? Well, superficially – yes, but if we dig a little deeper into Beethoven's history we may also discover a benign aspect to his deafness.

Ludwig van Beethoven had an unhappy childhood, primarily as a result of the incessant exploitation of his musical gift by his abusive, alcoholic father, who – while lying about his son's age – wanted to present him to the world as another Mozart-like child prodigy. Although as a young man he fell in love with a woman for whom he wrote "Fur Elise (in German the "u" has an umlaut)," he never got married and never had a family of his own. However, after his brother's untimely death, the composer immediately committed himself to raising his young orphaned nephew. It turned out that his nephew had an abrasive and rambunctious personality, or, at least, so Beethoven thought. Not surprisingly, over time, the presence of this nephew began to annoy the composer greatly, who – according to his friends and his landlady – wasn't exactly a "laid-back" personality himself. But he did take the promise made to his brother seriously, and never for a moment considered placing his nephew in an orphanage. In any case, this domestic situation must have interfered with his ability to concentrate on his compositions. Lesser men might have fallen back into inferior levels of performance, but not Beethoven, for in his case – the perfect music in his head simply "had to get out."

It is not unreasonable to speculate that it might well have been a superior sub-conscious wisdom, which – in response to the many unhappy events in his life as well as the immediate annoying outer stimuli – gradually began to close off the hearing path to the brain in order to shield his power of concentration. It must have been a subconscious action of course, because it's highly unlikely that Beethoven ever said to himself: "I think I should turn deaf to solve my problem." But what an incredible loss it would have been to the world if these later works of the composer had been composed at a mediocre level – or possibly not at all.

So, was the loss of hearing all bad, really? Surely, only simple-mindedness and shortsightedness would classify Beethoven's deafness as a straight "failure."

Now, somebody could correctly argue that there are better ways to "tune out" the conflicting sights and sounds of the world around us. But if these methods are not known to the individual, it is the same as if they don't exist. The deeper wis-dom "grabbed" what was available to the composer and wisely ignored what was only potentially (and therefore obviously useless) possible.

There are so many other examples of a similar nature that prove that being "cut off" from the "common" human experience by a handicap doesn't necessarily have to lead to an inferior existence at all. Could it be that one of the outstanding violinists of our time, Itzhak Perlman, who contracted polio at the age of four – de-priving him of the use of his legs – took proper advantage of his restrictions, such as not being able to engage in typical young boys' activities, by focusing on the mastery of the violin; and was the handicap not more an aid than a deterrent to his becoming a superb virtuoso on his instrument? Many, many people would gladly exchange such a "rich" life for a drab existence with the use of their legs!

And then there is Helen Keller, etc.

Of course, it would be an outright absurdity to ever call a handicap, or a disease a good thing in itself; and it is our absolute and unqualified right – or perhaps even duty – to try to eradicate diseases wherever possible. It's certainly not necessary to "buy into the belief" that we must suffer a little in some areas of life so that we may experience "the greater" in other areas. However, if we begin to look "past the end of our noses," so that a larger picture can emerge, we may discover that even those "incomplete human experiences" can potentially catapult us from the center (or even from the rear) into the more forward and far more enjoyable echelons of evolution – which is the region of greater self-actualization with its attractive awards of, among others, a strong sense of self-fulfillment and, yes, quite possibly greater finances.

Although it may be possible to speak of reaching Abraham Maslow's point of complete self-actualization (even though there is obviously no exact point of demarcation because it is still a relative designation) – it should be understood that there is no such thing as the final denouement in the evolution (or expansion) of consciousness process, because – judging by the past – that process should go on to infinity. Its upward spiraling probably never ends – because it is inconceivable that there is a point where discovery can go no further and learning and becoming has come to an end.

Again, is it all bad, really?

What to do when we don't understand it all

Now, somebody could say: "OK, fine, you've given us a few examples of handicaps and diseases that may have benign aspects to them. But what about the "bad gene diseases" that can cause infants to die young where there is obviously no opportunity "to take advantage of their restrictions." Isn't such a genetic defect a strike against model 1 and a compelling reason to add some weight to the side of the scales belonging to model 2?

That's certainly a fair question but the answer is still "No"! Why? Because at a juncture like this, where it may tempting – particularly for those who had a direct experience with such a disease – to condemn "the whole system," we have to go back to our standard – which is also the most disciplined scientific attitude – that dictates that we must suspend judgment or refrain from coming to a conclusion when we don't understand the causation behind a phenomenon.

An analogy may be helpful here. Suppose we go back to the days when we hadn't proven conclusively or to everyone's satisfaction that the earth is a spinning ball in space. Now, we're already at the point when all reasonable evidence tells us that the earth is indeed round, and then we suddenly run into a "puzzling snag" which seems difficult to reconcile with roundness. Do we then get up immediately and become "flat-earthers" again? "Not hardly," as John Wayne's movie character would have put it!

Of course, such an answer may still "not sit well" with the objector who might continue with: "Well, that's very convenient, so what does it take for you to begin adding weight to the side of model 2? The answer to that question is simple. We would have no choice but to add some weight to the side of the scales belonging to model 2 *if* the genetic complexity level was relatively low   low enough to be within the "reach" of short-armed Randomness – and the occurrence of these abnormalities (the frequency rate) anything other than rare. But the latter is rare and the former (see chapter 1 on DNA complexity) carries odds against model 2 that are, without any exaggeration whatsoever, of truly astronomical proportions.

If these examples of "failures" in individuals, that once might have been seen as evidence of "Life's failures," no longer compel the well-thinking mind to lean in the direction of model 2 (chapter 1), how do we make sense of the greatest of all "failures"? That is the subject material of our next chapter.

Not exactly a cheerful subject

There is a prevailing belief among members of the positive thinking community which tells them that they must only dwell on subjects of a cheerful, or life-affirming nature; and that discussions on topics that may be classified as "negative" should be "avoided like the dickens." Although I am largely in agreement with that philosophy and share the optimistic view of Life, the refusal to squarely look at the so-called dreadful aspects, or phenomena of the human condition keeps the corresponding – conscious or subconscious – fears unnecessarily in place. It is understanding, which means that we must spend some time studying the nature of these fears, and *not* avoidance – like the proverbial ostrich with its head in the sand – that can lead to their eradication.

As I have stated in the beginning of this chapter: "Aggression isn't bad per se …" – the same holds true of Fear. For example, one may lament about the fact that he or she suffers from acrophobia, or the fear of heights. But that is being unmindful of the fact that this fear is essentially a friend and not a mistake of Nature. Certainly, for unknown reasons the fear may have become excessive, but that doesn't deny the fact that it is fundamentally there to protect us.

There are members of a certain Indian tribe who are known to be free from the fear of heights. Some of them have been hired as steelworkers for the construction of skyscrapers, because they experienced no discomfort walking over steel-beams at great heights above the ground. But before you would say: "How wonderful …," consider the "flip side" of that comfort: more than a few have misstepped – because the absence of "that friend" resulted in carelessness – and I don't think they had much of an opportunity to enjoy the scenery on their exceedingly rapid return to earth.

Of course, there are fears that we can now well "do without." Thanatophobia, or the fear of death, which also has a definite – tied in with the self-preservation principle – function in the evolutionary process, may be one of them. So, there is no need whatsoever to approach it (the fear) and the phenomenon of death itself, as "the Great Enemy," or the last enemy to be conquered, for if it ever was beneficial at all then it cannot be considered evil. I have taken the opposite approach in the next two chapters.

Chapter Four

The question of Death and Evolution

There is little disagreement between reflective human beings of all sorts of backgrounds and persuasions about the fact that there remains one aspect of human existence that continues to baffle virtually all of us. The primary reason that questions of a philosophical nature first began to be asked in ancient times is that there is an end to the life of Man. In the course of history, an almost infinite number of words have been spoken and written to try to make sense of the reality of death. The question of interest here is: "Does death have a purpose (model 1, chapter 1) or is it just unfortunate and meaningless (model 2, chapter 1)"?

Before we begin to relate to that question, it must be stated, up front, that no special insights or revelations are available with which we can "paint" a picture of what the experience of the departure from this world will be like for a human being – ourselves included, of course.

Surely, the previous sentence must invite the reaction: "Well, if you are not going to give us any meaningful and direct answers, why even bother writing about this subject?" The answer to that question is a very direct one: "Because, even though direct and completely verifiable answers about the phenomenon of death, which are *not* based on a belief-system, do not exist – and no one can claim that they do – and no belief-system is being advanced in this writing, it is, nevertheless, possible – in spite of the absence of a "religious revelation" – to make (some) sense of the phenomenon of death through a very different approach, namely the utilization of the mathematical principle of extrapolation."

This answer doesn't imply that there is anything wrong with having a belief-system or holding "pre-knowledge beliefs" as such, because there are helpful beliefs "out there" capable of furnishing the human psyche with a substantial level of comfort in times of grief and sorrow. When a parent goes through the truly devastating experience of losing a child, the support that he or she may get from the belief – particularly when it is at the level of a conviction – of the "on-going-ness of life" and the possibility of a future reunification is indeed priceless. From a purely pragmatic standpoint alone, no case can ever be made against holding such a helpful belief.

However, at the same time it must be said that compared to knowledge, a belief is still not much more than the "proverbial sky-hook," which doesn't connect to anything. We all know that it is quite possible to "talk (particularly through repetition) somebody out of a belief," even though it is also true that the more deep-seated beliefs are much harder to dislodge. But we also realize that the average adult cannot be "talked out of the fact that $2 + 2 = 4$ (because I have used this simple mathematical equation earlier, I was tempted to use the "all-new" and "exciting" example of '$3 + 3 = 6$' for a change – but, no doubt, you would have caught on to my creative ploy. Oh well, …)." Why? Because, unlike with a belief, knowledge – having nothing whatsoever to do with the art of persuasion – allows us to quickly re-trace the steps of a verifiable principle, so that we can – with certainty – get back to what we once knew. This is the very reason why so many different beliefs and belief-systems are possible, because there are usually very few, if any, verifiable re-traceable steps (back to a solid foundation) in them.

In some countries in the Middle East, young men are told that Allah will supply them with forty-nine virgins (apparently having no say-so in the matter) in the hereafter as their reward for, what we in the West consider, a committal of atrocious acts. That's obviously much easier to do and less expensive than handing out a cash reward, because there are no verifiable re-traceable steps to any reliable base, or point of origin in that promise. Unfulfilled campaign promises may sometimes catch up with a politician, but this "virgins promise" requires no signature at delivery time.

The Art of seeing much more than the obvious

So, the all-important question here is: "Can we find any kind of 'advantage' in the fact that a life cycle has a definite time limitation"?

I can almost hear somebody say: "Man, you've got to be joking, because the concept of death and the idea of advantage do not go together." A skeptic once said to a church-goer: "If you guys can ever make sense of dying, I'll come and join your religion, but since I know for a fact that you can't, I'll be an atheist for the rest of my life."

Now, if the individual – human or animal – were an "arrived" manifestation of total perfection without any need or possibility for further improvement, in other words, an instantaneous perfect creation at all levels (physiologically, mentally, etc.), then the limitation of a life cycle makes no sense whatsoever. However, if he (generically) is still evolving, perhaps this limitation does have an advantage or may even be necessary.

To explain this "life cycle limitation advantage," we must go back to the last chapter where it was demonstrated with examples that "life" has a twofold purpose: learning (to live better) and enjoying (both being equally important). Now, this kind of learning culminating into a personal transformation (from aggressive to tolerant and caring, from "doormat" to being assertive, from laborer to artist or businessman, or – in the eighteenth and nineteenth century – from slave trader to abolitionist, etc.) requires flexibility and adaptability as well as a willingness and readiness to change (for the better of course); and if one is ready to continue to grow, these "virtues" must remain in place during one's lifetime. Unfortunately, when we take inventory of these particular virtues among our elderly population, we will notice that, so often, the very opposite is happening in old age, and that some sort of "mind crystallization or petrifaction process" – the grouch becoming grouchier, etc. – is well on its way.

The Greeks identified the Great Law of life well: "Grow or perish." So when there is no more growing, its only alternative – because it is an immutable law – *must* "kick in."

If the South would, today, be populated by a majority of 200-year-old white individuals and there wouldn't have been a Civil War, it is not an outrageous stretch to the imagination to think that slavery might still be in existence today – or what is a virtual certainty, that blacks wouldn't have been able to rise to the unquestionably higher – although not yet ideal – status in society that they now enjoy in the twenty-first century. Didn't the slave owners choose the awful slaughtering of their sons on the battlefield *over* the relinquishment of their comfortable-to-them-but-miserable-to-others "life style"? The "life cycle limitation" – where not only "old bodies" but also "old minds" are taken out of circulation – is a fast(er) way for this evolution of consciousness process to overcome the – sometimes extreme – resistance to constructive societal changes.

It is not hard to come up with numerous examples of the entrenched attitude in Man. Not long ago in a Second World War documentary, an old SS soldier talked unashamedly as if the Third Reich were an imminent reality.

So, because the "generational turn over" facilitates the faster-evolving process – in the biological evolutionary process (insects evolving rapidly in a laboratory) as well as in the evolution of consciousness phase – have we now not identified at least one advantage of "the life cycle limitation" or Nature's "recycling process"?

Now, if society benefits from the "life cycle limitation" and the "generational turn over" of its members, it is not unreasonable to theorize – with the support of the principle of extrapolation – that the individual can also personally benefit (only under model 1 of course) from this recycling process, because he is – albeit without his permission – "mercifully 'yanked' out of this petrifaction process." Surely, we cannot – without the ability to look around "that corner" – absolutely confirm that, but the hypothesis is reasonable. Again, Nature has ordained that growth is mandatory and not optional!

It must be stated here that extrapolation doesn't furnish us with an ironclad proof, but it does tell us that the idea is highly plausible.

So, here we have the two "growth advantages" of a limited lifespan. The first one generates a quicker evolution of consciousness benefit to society and the second one engenders the "becoming un-stuck" benefit to the individual (of course this is only true if there is "continuity," or "a future life").

The third advantage: the push to grow on the heels of so-called misfortune

The third advantage puts – or actually forces – a "second" generation onto the road to greater self-reliance after the "departure" of the "first" generation, because then the "younger" suddenly becomes the "elder." Because aging is "supposed to produce greater wisdom" – which certainly isn't always the case (there are a lot of old fools around) – the human community, even in primitive societies, has bestowed the "right to speak with authority" on its "chronologically most mature members." The young man had to "shut up" until the old man had left the hut – but once the latter left through the chimney instead of a door, then ….

A middle-aged woman ran her own prosperous business with superb executive skills. However, when her mother died, she became – lasting long past the normal grieving period – a total "emotional wreck" and began to look to alcohol for comfort. Even though she had matured successfully in most areas in life, there remained one area where full (or a fuller) psychological maturity was never reached, namely, outgrowing the emotional dependence on a parent (in this instance a mother).

Emotional dependence is different from "feeling a great love for" a mother or father, or enjoying being in their company. It's also different from "seeking advice," or "appreciating their input," when that is strictly done at the "information gathering level." Now, Nature (or a Cosmic Wisdom) has decided that at some point the "emotional training wheels" *have to come off*, even if that leads to a period of bewilderment for the moved-out-of-their-comfort-zone human beings. At the proper time, the young eagle is forced out of its nest by the parents, because they know by instinct when their offspring is ready to take to the sky. So we are – in the absence of personal initiative – sometimes forcibly placed on the road to greater self-reliance. The unprepared daughter had her source of emotional dependence stripped away from her in that manner.

However, just because we're put or "dumped" on the road to self-reliance, doesn't mean that everyone begins to travel in the direction of that lofty destination immediately. No, the human mind is crafty and resourceful, so she decided – probably at an unconscious level – to delay the journey and pull back into her comfort zone by substituting her one dependence for another – the bottle. But that destiny – the goal of greater self-reliance – *has to be* reached sometime, so maybe even the dreaded "grim reaper" helps us in some way in the long run.

So, death is only a "totally bad mistake" to those individuals who insist that life should only be about enjoying it and "let's forget about all this evolving stuff (the other half)." All the "kicking and screaming" that we may do along the way doesn't allow us in any way to step out of the evolution of consciousness process, because we have absolutely no power whatsoever to alter universal principles. We can only work *with* them. If this is not done voluntarily, then sooner or later it'll be forced upon us; so it is pure shrewdness and the *right* kind of selfishness to do as much growing into the *right* kind of self-reliance as we can while we're here, because stagnation spells misery. Then perhaps at some point in our personal evolution these "cycles" aren't necessary anymore.

What about if the prophet was right on this one?

There is a rumor "out there," that according to an ancient prediction there will only be two more Popes after John Paul the Second, the current Head of the Catholic Church. I will readily admit that I, being a person who is somewhat interested in – but definitely not preoccupied with – prophecies, have not researched this prediction personally. However, if this prophecy were to materialize – and, of course, as it stands right now this is all nothing but pure speculation (could the pedophile-priest-sexual-abuse crises "do it in" eventually?) – then we would have, at a societal level, a very similar situation or circumstance that the above-mentioned business woman, who lost her mother, had to learn to deal with.

If an excessive leaning on the so-called wisdom of others is not "a good thing from a standpoint of the evolution of consciousness" and is therefore not "permitted to go on forever," then we shouldn't be surprised that at some point individuals, or even large groups of individuals, may be "forced" to switch their comfortable loyalty from a "spoon-feeding Spiritual Mother," who will gladly inform the faithful what is permissible to believe, to the "initially very scary idea of learning to trust and appreciate one's own capacity to reason about the core questions of life." If this force to switch from external to internal reliance is not thrust upon the "required to evolve" individual, or the group, as abruptly as the prediction suggests, the transfer still has to be completed at some point. According to the earlier stated Law of Life "Grow or perish": "Come in – it must." If not sooner then a little later.

The following observation may not be pleasing to some people, but there is little doubt that more evolved, or philosophically emancipated human beings rarely spend much time in church pews – although they may be found in an audience of a more enlightened speaker or teacher – waiting for someone else to tell them how to keep the "facts of Man and the Cosmos *straight* according to church doctrine." It takes some daring – and a love for Truth which is far greater than the love of repose or mental inertia – to become a common-sense-oriented heretic. Of course, the key words in the previous sentence are "common sense," which would prevent one from foolishly marching in a certain parade (figuratively) just because the nice and persuasive leader sold him or her a "good" story – or, more accurately, "a bill of goods."

Many decent and otherwise intelligent churchmembers will sing songs like: "…His Truth is marching on" without realizing that Truth *does indeed* march on, but that It often *keeps on* marching well past the point where they think it should have stopped; and that it is their love of repose – which translates into: "sitting on your evolution of consciousness duff and refusing to get up" – and their fear of letting go of incomprehensible indoctrination that prevents them from seeing that. Another well-known song says: "…and we'll understand it better by and by," but when a real and actual opportunity for greater understanding presents itself, that same solid indoctrination will make them say: "No thank you." That the failure to recognize Truth is not at all uncommon is exemplified in the story of the "two men (disciples – and therefore 'insiders') walking along the road to Emmaus …" in the New Testament of the Judeo-Christian Bible (Luke 24:13) as well as in many other passages (I trust that the interested reader can research that story without my elaboration).

However, paradoxically – yes, we might as well get used to the idea that there are a great many paradoxes in life – it must also be stated that the process of subjecting one's worldviews to the opinion of so-called authorities may (temporarily) have been desirable, or may even have been necessary at a much earlier stage of individual growth. The fact that there are many, many stages of growth, which all require different methods and approaches to assist the individual during his internal pilgrimage to the next level, has been completely overlooked by most of the early religious philosophers, such as for example, the rebellious-"sinner"-in-his-youth-before-his-conversion St. Augustine, and – I'm sorry having to state it like that – his "non-thinking" clergy followers throughout the ages. One size doesn't fit all! We certainly do not all need the same messages and reminders to spur us on to the next station on our journey. Yes, I have little doubt that the slave trader who became an abolitionist was indeed tremendously helped by the realization of the wretchedness of his pre-conversion mentality, and substantially benefited from the "sin (an ancient archery term for missing the target or mark) and salvation message," which some preacher must have communicated to him. The bottom-line is that it got the slave trader through the passage he needed to go through – and so, the pragmatic value of "the salvation story" cannot be argued. However, that doesn't – in any way whatsoever – mean that Albert Schweitzer was a wretch, and should have considered himself a sinner to be saved by embosoming a "supposedly universal" incomprehensible nail-somebody-to-a-tree doctrine in which this highly evolved man didn't believe. The Universe is far more sophisticated than that – and it behooves Man to take on some of that sophistication.

Yes, the slave trader unquestionably needed to take that sharp turn in the road – but the great doctor of Lambarene, Africa, didn't – because he wasn't on that same road. None of us would say that once hydrogen has "burned" into water, that the water is still hydrogen and oxygen – so, "why the heck" do some folks insist that once a human being has successfully made that turn, we still have to identify him or her with, or by the former self. A bit of thoughtlessness, maybe? Or stupidity?!?!

One shining example of the stupidity of that the-same-rule-for-everybody philosophy involves the enrollment of a Western Canadian student, who had been an "Anglophone (an English speaking Canadian)" all of her life, in a Ph.D. program at a university here in the States. The young lady, who had already earned a Master's degree in psychology in (English-speaking) Canada, was required to take a for-immigrants-designed-English-as-a-second-language class simply because she only had foreign credentials. Ssshhheeeee ..., was she ever having a good time – practicing pronunciations of basic words and all of that good stuff – in that class, probably tutored by a recent immigrant from the other side of the Pacific. And then the highly bureaucratic, licking-his-chops-while-staring-at-the-unnecessary-additional-tuition-fee administrator who easily could have exempted ..., but didn't want to sign off on .... I don't know about you, but I must admit that I consider this only-to-be-found-in-the-human-kingdom story hilarious too.

However, I do suspect that she integrated well after the "salvation" of that class.

We don't live in our parental home forever

Obviously, we must first be introduced to certain facts – possibly, but not necessarily, in a church environment – before we can begin to question them. Nonetheless, one doesn't live in the parental home forever. You've heard it before: "The only thing that is permanent in life is change." The inflexible ones will get hurt by it, the flexible ones will thrive on it. You keep the grains and discard the chaff.

Unlike the often-quoted philosopher, Lord Bertrand Russell, who didn't see any need to give the still-venerated-by-the-millions old religious institutions much credit, I must say, in their defense, that – in spite of their, in my opinion, question-able doctrines, dogmas and political views – they have been the guardians of valu-able philosophical writings for thousands of years.

If model 1 (chapter 1) is the correct model of the Universe, then it is reason-able to assume that an innate Wisdom can "take over" and fill the void that may have been created by whatever societal event or development. There is no reason to believe that we will be left "hanging in a vacuum" because that would deny the presence of Intelligence behind the forward step. Einstein, who was manifestly disinterested in organized religion, nevertheless, attributed his hunches and in-sights to a Cosmic Intelligence.

Once you have made the transition successfully and you are comfortable in that new habit of thinking, you then have become your own Pope – albeit with-out gold-embroidered white robe and miter. That's not being arrogant – because you grant that status to all your fellow human beings – but it is simply the natural progression in your own personal evolution of consciousness. It's like the expres-sion: "Every man a king, but no one wears a crown." The just-mentioned "giant among men," Albert Einstein, didn't see any need to dress up and decorate "the outer shell" like a Christmas tree. He was frequently seen wearing a sweater, with his hair all wild, and without any socks on. So, if you want to be "a somebody," don't wear a suggesting-great-authority, sixteenth-century robe …, just take your socks off ….

This idea of accepting personal responsibility for what goes on in the "inner Man" has a parallel at the physical level. Many people are now waking up to the fact that taking initiatives concerning their own health, such as not pulling into the parking lot under the "yellow arches" every time the bodily fuel gauge reads "empty," or not indiscriminately swallowing every pill that the beholden-to-the-pharmaceutical-industry man or woman with the stethoscope will prescribe, is all part of the process of becoming more aware.

Now, if this prophesied only-two-more-Popes-after-the-current-one "shortly coming up" event were to occur – and, believe me, I'm not biting my fingernails in anticipation – one cannot blame the individual who made the prediction, or the author who wrote about it, because neither one has, or had, any power whatsoever to cause it to happen. That power, if it exists at all, is universal and by no means individual. There is no point in killing the messenger – as was done "in the days of old"!

The fourth advantage

Another – if properly viewed – benign aspect of the lifespan limitation is that it can potentially create a "good" sense of urgency in Man. I used quotation marks with the word "good," because it is of the utmost importance to see when and how this sense of urgency is benign, and when it is not, or not entirely so.

Our eighteenth President and magnanimous-in-victory Civil War Union Army commander, Ulysses S. Grant, who – notwithstanding the high esteem he enjoyed with his peers and the general public – was swindled out of a fortune by his two greedy Wall Street banking firm partners in later years, hurriedly wrote and completed his – generally considered well-written – memoirs while losing his last battle with throat cancer, in order to provide much needed finances for his family at a time when Presidential pensions did not yet exist.

There is a definite parallel between Grant's writing and the frantic attempt by the fatally-ill Wolfgang Amadeus Mozart – who also desperately needed money – to complete the commissioned composition of his Requiem; this unfinished work was completed by the composer Franz Xaver Suszmayer. In Mozart's case there seems to be some historical evidence that his ignoring of "Nature's proper rest requirement" aggravated his illness and accelerated the dying process. These are obviously examples of the wrong kind of urgency; they are not to be emulated.

On the other hand, if we travel to another part of the world, with a culture totally different from ours, such as India for example, we meet human beings whose native ambitions and aspirations have either largely, or completely been purged from the mind. This "too relaxed mindset" presumably originates from the strongly-held reincarnationist's viewpoint: "If not now, then in the next lifecycle …, so what's the hurry?"

Well, yes, the relaxed state of mind may be "a good thing" – but apathy certainly is not. A well-balanced sense of non-hurried urgency (our fourth advantage) can nudge us toward greater self-expression. Walt Whitman said: "For far too long have I slumbered upon myself."

If Man were to receive the Celestial gift to live a thousand years on this planet, there is a high likelihood that the vast majority of men and women – given their current evolutionary level – would "waste (putting off worthwhile projects and neglecting the development and nurturing of talents, etc.)" much of the additional nine hundred plus years as a result of the still-very-much-present, already discussed conservation-of-energy mind pattern. Give Joe Average two days to complete a certain "do-able" project with threats of dire consequences if doesn't get it done, and he will most likely finish it before the deadline. Give him three weeks without threats for the same project and he will – with the most reasonable sounding excuses – plead for more time.

The IRS gives the taxpayer three and half months into the year to file a tax return, which in most instances can be completed in a few hours or less. But notice the long line of cars at the post office – even in our present-day of convenient electronic filing – just before midnight on the 15th of April. How do I know that so well? Yes, ahum, uh …, I have impatiently counted the cars ahead of me from my own jalopy, but, I am proud to announce that I, without any help from Procrastinators Anonymous, have been a "reformed" or "transformed" – applying this evolution of consciousness "thing" – man for decades. Believe me, the "lousy" non-Espresso coffee they handed out isn't worth the wait!

So, again – on balance – this life-cycle limitation isn't "all bad" at our present evolutionary stage.

The fifth advantage

Another interesting advantage of the life-cycle limitation is that Man must travel through the seasons of life that have their own distinctive charms. If in our later years we would still have the same physical strength, endurance, and agility that we enjoyed in our late teens and early twenties, we might be tempted to continue to live exclusively in the physically-oriented lifestyle of our youth and ignore all the wonderful "adventures" of the following levels or stages.

Philo Judaes, the Alexandrian Jewish philosopher, said: "In our youth we practice the athletics of the body, age is for the athletics of the mind." I'm sure he wrote that in his later years.

The phenomenon that young boys (or even young men) admire or idolize, for example, a Mike Tyson, and rarely an Albert Einstein, is an extension of Philo's wisdom.

Many of us are familiar with the aging actress who cannot and will not accept the fact that her looks – despite the face-lifts and all the other cosmetic manipulations – are permanently and irreversibly slipping out of the gorgeous category. Henry Victor Morgan called himself "flaming youth" in later years, but upon closer scrutiny we must conclude that he was only badly kidding himself. Am I suggesting that we may not, or should not, attempt to rejuvenate the body or the physical appearance whenever possible? Not at all! What I am saying is that Life offers many gifts and not just the one. There are probably some less temporal ones among them, so why focus exclusively on the one?

There is an old saying "as one door closes, another one opens," so why remain standing in the door opening when the closing door is about to hit you in the face? Almost everyone who courageously walked into the next room of life will attest to the fact that stepping into the new experience, with wisdom, has turned out to be every bit as rewarding as the former experience; and you still own the latter's treasured memory.

Even though I – being past middle age according to the opinionated calendar – am very much interested in concepts like "healthy aging," I do not feel cheated by the fact that it is no longer wisdom to plan my weekends around downhill skiing trips, as I did thirty years ago. I am presently having a "blast" being in this not-so-hard-on-the-knees writing adventure, without being overly concerned about prospects for publication, and without agonizing over whether or not I have – while striving – achieved ultimate perfection in eloquence and grammar. The only sure way to avoid making a mistake (in public) is to do nothing – and that is the most tragic mistake of all. And so, with all my (and everybody else's) compulsive, non-constructive, nitpicking critics – literary and otherwise – I share the following profound, loving, and highly therapeutic thought, which came to me in a moment of supreme enlightenment: "Pooh pooh!"

If you, the reader, feel cheated in the above-mentioned way, and are overly distraught over the loss of your youthful sprightliness and stamina, you always have the option to reflect deeply on how your unique later-in-life "athletics-of-the-mind" adventures can come into focus. They'll "come in" (possibly in great detail) as sure as the sun rises – but without that sarcastic profundity that just rolled from my pen.

Utopia?

Are these five advantages and the obvious advantage number six: the very benign suffering-limiting aspect of death – which doesn't require much elaboration – real, or are they just Utopia, in other words: a trying-to-make-the-best-of-a-bad-situation upbeat attitude which accomplishes nothing of substance? I'll leave that to the reader to answer.

Applied common sense to overcome thanatophobia

Because thanatophobia, or the fear of dying is a looming reality in the human psyche, it may be useful to see if common sense reasoning can help us allay that foreboding uncertainty (at least to some degree). Without any so-called religious insights, we can still decide that:

1) a future life may not exist at all (model 2) ------ if this is true, then there is nothing to worry about because even though this is not a happy prospect, the concept of non-existence shouldn't frighten anyone.

2) a future life will be as good as our present one ------ if this is true, then there is nothing to worry about.

3) a future life will be better than our present one ------ if this is true, there is certainly nothing to worry about, and it is a happy prospect.

4) a future life is worse than our present one ------ if this is true, we do have something to worry about – but – we can get rid of that worry, right here, in the next few paragraphs or so.

The concept of "intense suffering and misery beyond the grave" comes from the religious corner, of course. Artists from the Middle Ages, such as Hieronymus Bosch, Lucas van Leyden, and a little later, Michelangelo Buonarroti, have left Mankind vivid and graphic images in their paintings of the "the Last Judgment" and "the condemned souls descending into hell, or the lake of fire," concepts taken from the Judeo-Christian Bible. Even in the twenty-first century, preachers are still routinely making reservations for the non-believer in the "unmentionable" hot place. Because we cannot send a news reporter down there to verify the story, it can perpetuate itself endlessly.

Consistent with our thus far stated perspective, no alternative belief will be placed in opposition to the belief in a future hellish experience, which might then have been followed by a statement, such as: "Now this is my version, and because it makes a lot of sense, you should really believe that instead." There is no point in substituting one irrational belief for another. If we cannot deal with this concept at a purely rational (non-belief) level, I would then prefer to "scrap the whole mess." But we *can* deal with it at that level, and I believe quite effectively.

Now, the concept of eternal damnation is – for two reasons – not anywhere near as innocent or harmless as it may appear.

Years ago, a twelve-year-old son of a friend of the family was invited to go to church by one of his friends. His favorite uncle had died a few months prior to that invitation. A few days after that church attendance, he woke up in the middle of the night with a loud scream. When his mother asked him what was wrong, he told her that he had a nightmare in which he saw his uncle in hell being chased by devils with pitchforks. Apparently the devout in that church had wasted no time trying to indoctrinate the young boy.

In addition to this relatively quick subconscious feedback in the twelve-year-old boy, there is also the possibility of a delayed – occasionally taking several decades – reaction or feedback, which may, as a result of initial incorrect processing, still create a fair amount of turmoil in the human mind after that many years. A great-uncle of mine was raised in a strict Catholic family. But by the time he reached adulthood he had turned his back, not only on Catholicism, but also on religion in general. Although, I don't think he ever became an atheist or an agnostic, he was definitely an irreligious man with an amazing arsenal of religious jokes. However, during a major illness in his twilight years – from which he briefly recovered – it became quite clear that all the nasty concepts and images of his early upbringing were still making their ugly presence felt with thoughts and emotions of fear and foreboding. As a young man he may have been shown copies of the paintings of Hieronymus Bosch as part of his – "warning the sinner not to go astray" – religious education. These paintings may provide the more enlightened mind with a good laugh or two, but they are capable of scaring the wits out of the young and highly impressionable type of mind. I think they are quite amusing or even hilarious now, but when I first saw them as a young teenager, my reaction was that of shock bordering on horror.

Undoubtedly, my great-uncle had expected these religious images to have disappeared into oblivion after more than half a century of "laughing them off."

Well, the method of "ignoring," or "joking about it," will apparently not always get rid of the effects of earlier indoctrination. The conscious memories of the stories of heaven and hell may have faded and moved out of sight over the years, but they ended up in the back of a filing cabinet and not in the trash can. It all floated back to the surface when the time of death came closer.

So, how does one process and dispose of those unwanted beliefs correctly?

This is my method:

1) Recognize that this present moment is a good time to assign this concept – without apologizing to yourself or anyone else – to the category of "pathetic imaginary trash," where it belongs.

2) Recognize how those beliefs originated – you'll "figure it out."

3) Recognize that the indoctrinating adults did not have a "lock on Truth" – or using a popular term: "the insider's scoop" – and were either in their ignorance presenting an outright falsehood, or a poor interpretation of something that somebody wrote a long time ago.

4) Recognize that those "dumb" beliefs always fall apart under the "microscope."

5) Recognize that fear loses its grip and power in the well-ordered, more enlightened mind.

6) Recognize that a Cosmic Intelligence could not possibly be the barbarian that some people have created "in *their* image and in *their* likeness."

Let's demolish the eternal damnation concept, right now!

We could use an illustration in which a-throw-the-book-at-him judge condemns the already earlier used villain (I might as well put him to work here, so he was at least good for something), Adolf Hitler, to the incredibly severe sentence of one million years of "roasting" for every death that he was responsible for in his life-time. That is – pardon my apropos vulgarity – a "hell of a lot" more suffering than what any of his victims had to go through. But why worry – according to this judge and the sentiment of millions – the "s.o.b. (son of Beelzebub) had it coming." History tells us that about twenty million people lost their lives as a result of this man's ruthless decisions and actions.

If the reader allows me to indulge in a little digression here, it must be borne in mind that had the man been surrounded by uncompromisingly principled and non-opportunistic people, such as the philosopher Socrates, Bertrand Russell, Mahatma Ghandi, Winston Churchill, the German theologian Dietrich Bonhoeffer – who was executed by the Nazis, Mother Theresa, Dr. Martin Luther King Jr., and numerous other great men and women in history, instead of the porky "art collector," Hermann Goering, the incurable liar, Joseph Goebells, and all the other henchmen without a conscience, as well as the culpable segment of the German population of that time – then his megalomania and his perverse oratory would probably have taken him to an insane asylum and not into our history books. But if he would have become one of the committed "unfortunates," because, without the aid of his despicable cronies, he never had a chance to create wholesale misery for others, should we then be generous and place the poor guy in the slightly more forward echelons of evolution. Of course not! Unexpressed ignorance is still ignorance; and unexpressed backwardness of any kind is still backwardness. Isn't it the consciousness of Man rather than his actions that counts most of all in terms of such placement? But here I will stop with my digression.

Let's get back to the point where we need to bring out the calculator. Twenty million times one million equals twenty trillion. So, Adolf Hitler – since he can't die during all that time – can look forward to being released in the year 20,000,000,001,945. That, dear believer, is *still* finite time and *not* everlasting or infinite time. Now, if Hitler is kept in that place of torment past that point in time, then the judge, God (the God of Love?), is worse than Hitler. As the saying goes: "Try to put that in your pipe and smoke it." It simply doesn't add up. That would be "two eyes for an eye" instead of "an eye for an eye," and no one can argue that that wouldn't constitute a gross injustice. Twenty trillion years of intense pain and suffering – give me a break!

But we still see bumper stickers reading: "Where will you spend all of eternity?" Yes, the most reverend Ignoramus Whop-de-do, who seems to be incapable of seeing past the end of his nose, doesn't give up easily. He is with us yet!

An inaccurate use of one word

Linguistic research reveals that the ancient scribes from the Middle East, the area where these concepts presumably originated, did not make a clear distinction between the words "indefinitely" and "eternally," the way we do. Writing the words: "the way we do," reminded me that it isn't even everybody's way in the West. For how often do we hear the word "indefinitely" used around us in this part of the world when the speaker should have reached for the word "permanently"? Indefinitely can mean ten seconds, twenty hours, or thirty years. It points to an undefined or undetermined period of time.

Now, the word "indefinitely" fits in precisely with the way Nature punishes. Yes, Nature punishes Man, but never vindictively, and never beyond the point of recognition of the underlying causes of the problem – if that recognition is linked to the awareness of a principle(s) with which we can change the train of causation, followed by the required action on our part to apply the principle in the way it works.

Let's pull out another example here. A math-student wrestles with a problem in his or her homework assignment. After an hour or two, he or she is getting more and more frustrated (Nature's punishment telling us that we are on the wrong track) with the problem because nothing seems to work. Finally the light goes on inside the student's head, and he or she flips back to the instruction page where the problem solving principles are explained. Aaaaahhh, enlightenment: "I've been working with the wrong formula for all that time, but now I've got it." Euphoria sets in and the frustration disappears in the "twinkling of an eye." So, Nature's punishment, the frustration in this instance, only remains in place during the period of non-recognition, or ignorance, and doesn't last one infinitesimal moment beyond that point. You see, only human punishment is of the variety that punishes FOR wrong actions and unacceptable behavior, etc., but in the natural order of things we are only punished BY our mistakes and never FOR them. After solving the mathematical problem successfully, we don't expect the ghost of Pythagoras or Archimedes (some of the mathematical "fathers") to appear saying: "Sorry kid, we're temporarily out of our supply of euphoria packages...." No, our return to "heaven (euphoria – or simply feeling good)" – at the biological level: "the serotonin level sky-rocketing" – is instantaneous.

Granted, not every human problem is as simple and "straight-forward" as our example, nevertheless, this above-mentioned principle holds true even with problems of far greater complexity.

So, if we go back to the principle of extrapolation, which says: that which applies inside the known area will most likely also apply in the unknown area – which is far more reliable than projections of "sky-hook beliefs" – then we may rightly assume that the likelihood is "pretty good" that there is no vindictiveness "on the other side," and that new opportunities to change will at some point become available to the individual. However, since there is almost always a kernel of truth in any kind of belief, it is also quite likely that the "self-punishing, deeply entrenched dark and ugly attitudes in the human mind" and their corresponding self-inflicted (and therefore not superimposed) misery and suffering, won't be any easier to dislodge (indefinite punishment) on the other side than it is on this side. However, the belief in everlasting damnation is still pure poppycock!

An objection may be voiced here: "Aren't folks entitled to be left alone in their beliefs?"

Well, yes and no! Only the helpful and harmless ones, such as the above-mentioned "hope of reunification" belief, are completely off-limits – BUT wasn't the 12 year old boy with his philosophically immature (and thus defenseless) mind – no "security forces" available to guard the "compound," or no sentinel at the gate – entitled to a night of uninterrupted peaceful sleep? Where are the priorities here?

I certainly acknowledge the fact that it is almost never appropriate or "right" to verbally – let alone physically – attack another person (I'll leave the "split infinitive" stand here, because it sounds better – and, besides, all my old teachers are long gone anyway!). However, it is – as I have already unapologetically done earlier – unquestionably legitimate to expose the absurdity of an irrational belief, because in truth nobody is "married" to his or her beliefs or belief-system, although, for the time being, he or she may think of it – as a result of potent conditioning (like brainwashing) – as another until-death-do-us-part relationship. I, myself, have walked away from the belief-system of my youth – in spite of early pledges of loyalty and a profession of faith – when it no longer made sense to me; and that act did not result in a reward-for-the-unfaithful lightning strike; and left me no scars or bruises. Why does the ecclesiastic become more than a little perturbed with words like these? Probably not primarily because he is a staunch defender of the faith – although he will undoubtedly explain it to himself that way – but more likely because it may release his "wealth – meaning his flock" from their non-thinking captivity, never to return. For has anyone ever heard of the venturous one in the flock – who is capable of digesting the more nutritious food of the meadows beyond – going back to the former confinement voluntarily after escaping, or being set free? So, the Voltaire type of man, who noisily rattles the lock on the gate of the compound before opening it, has done "the flock" a great favor in the long run, because the stagnant water in the pond inside the compound was – without the flock being aware of it – becoming poisonous.

Of course, nobody should labor under the illusion that his single stroke of the pen will go far in expurgating these bordering-on-the-obscene beliefs in the mind of Man. However, such a single stroke may help the more adaptable individual in disarming and dismantling such a belief, so that its parts can be dumped into the recycle bin, or better yet, into the incinerator.

A curious concern for the individual in evolution

In one of the documentary-type Nature shows on television, a gnu or wildebeest lay dying as the result of a lion attack. But for some unknown reason the lions decide to forego their "dinner." The commentator said that the fatally wounded animal, appearing relatively relaxed, was in a state of shock. According to modern biological wisdom, the brain produces the natural painkiller(s) "endorphin(s)" in that state.

It is exceedingly difficult, if not downright impossible, to explain the presence of that "natural drug" in terms of the natural selection process. Now, the reader must remember that this statement is written in an ambience that isn't in any way "hostile" to the concept of natural selection (see chapter 1) – quite the contrary

– because that concept goes a long way in explaining most of the evolutionary adaptations and advancements. On the other hand, just because natural selection seems to be the best way to account for these two, there is no reason to go overboard and totally prematurely crown the phenomenon as the absolute ruler of *all* organical or biological development. The cautious non-absolutist – the true scientist – prefers and adopts the I'll-put-the-egg-on-my-breakfast-plate-instead-of-on-my-face resolution and attitude.

So, why is the natural selection interpretation virtually untenable here? Wounded or dying animals with the "engaged endorphins," like our wildebeest, do not have the greater advantage – quite the opposite – of passing on their genes (their capacity to mate being reduced or eliminated) to the next generation compared to their healthy "non-mutant" cousins. In other words, the presence of these making-the-individual-comfortable painkillers does absolutely nothing to trigger a new survival-of-the-fittest process. The above-mentioned "euphoria" natural brain drug "serotonin" is equally hard to explain in that way.

Randomness (model 2) cannot produce such a "caring phenomenon" in evolution.

Assistance to shipwreck victims

Another difficult to "explain away" (from the natural-selection-only standpoint) phenomenon is the benign behavior and even helpful assistance towards members of other species, exhibited by highly evolved animals.

Years ago, I read the story of a group of dolphins rescuing certain-to-drown victims of a shipwreck. They swam in a special formation: one under the exhausted human swimmer to keep him at the surface, most of the others flanking him on each side to guard against shark attacks, while one swims in front – apparently aware (resulting from their astonishing intelligence) that the swimmer is not a member of one of the marine species – to lead the group to nearby land, where they assisted the man in the "beaching process."

Now, granted, this may only be a nice story – I wasn't there, so I cannot vouch for its accuracy. However, from what we have learned of their behavior in captivity tells us that they are both capable and inclined to do something like that. When I watch these delightful "fellow beings," I often wonder if they are "playing" for the sole purpose of practicing (aside from the training by humans) and maintaining their survival skills – which is, no doubt, what many of the evolutionary biologists insist on – or if they have taken it one step further and are playing for the sheer fun, or joy of the "playing itself." And is the hold-on-to-my-flippers-and-we'll-both-have-a-fun-ride attitude in the dolphin, or the orca really merely a product of some accidentally-hitting-it-right rearrangement of a few brain connections?

Another example of this sort of highly intelligent kindness in animals was displayed – filmed under water by divers and therefore not just merely "cute" fiction – by a whale who raised his flipper at one point to avoid hitting one of the, compared to the enormous size of the whale, tiny divers. The whale could easily have – in the way we deal with annoying insects – "swatted the man like a fly." Such a benign action obviously "transcends" the survival of the fittest principle, because there is nothing in it that directly benefited the gentle giant (its extraordinarily powerful flipper wouldn't have been hurt by the collision in any way), or the species to which it belongs, in terms of survival enhancement. It is safe to say that, whatever it is, we are certainly not dealing with a simple, run-of-the-mill brain mutation in that class of whale species that prevented the diver from being knocked unconscious, or worse.

I am so happy to know that modern women go to Weight Watchers or Jenny Craig to make an adjustment to their figure, instead of strapping on a, made from a whale's baleen plates, old-fashioned, I-must-hide-my-avoirdupois corset. In spite of all the pessimism about him, Homo sapiens does make a step in the right direction every once in a while.

And then there is the story, among quite a few others of a similar nature, about the not-a-family-pet dog that curled up to a lost in the woods young child, staying with it until the child was found, just to keep it warm as well as protecting it. Oh what curious and warm-hearted intelligence, which only the trapped-in-his-philosophical-foxhole reductionist could attribute to the purely accidental evolutionary advancement from the recently discovered Eve, or the common ancestor of the animal kingdom: the sponge.

It should be obvious by now that no matter what phenomenon we are looking at, if we only dig a little deeper, we will begin to see the greater tenability of model 1 (chapter1).

Chapter Five

The question of Immortality

Most adult human beings from every corner of the earth have asked themselves the question at some point: "Will there be a future life for me and my loved ones?"

Oddly enough, there are some people who – at least at a conscious mind level – claim that they are "quite OK or even happy" with their belief that it all ends with the cessation of the brain waves.

A man in his early twenties walked into a church office and asked the minister if he would be willing to perform the wedding ceremony for him and his girlfriend. This request was followed by: "But I want you to know that I don't believe in any of this religious stuff, because I consider myself an atheist, and I know that there is nothing more than physicality. I know it's a bleak picture, but that's the way it is and I don't want you to try to talk me out of it." While he made those remarks he had a broad smile on his face and a gleam in his eye. So, naturally, the curious minister inquired about the reason for the church ceremony, to which he replied: "Only because she wants it."

That must be considered one of the most perplexing psychological phenomena: the smile and gleam while one talks about the bleak picture, because – at least theoretically – the atheistic belief in itself cannot give the adherent any cause for cheerfulness. Such a level of buoyancy, when one had expected no more than dispassionateness, tells us that there are two categories of atheists: "the diehard atheist" and the "reluctant atheist." The diehard atheist – being a devoted cheerleader for model 2 (chapter 1) – doesn't "want it any other way." One sometimes gets the impression that he would sooner undergo (a physical) open heart surgery than (a mental) change of heart surgery. Such an extreme resistance to a reality paradigm shift – in spite of an abundance of logically-based, solid evidence that mandates the consideration (not necessarily the acceptance) of such a shift – occurs in circles of orthodox science as well as in orthodox religion. However, the reluctant atheist – many in this category lean much more towards agnosticism than atheism – doesn't necessarily prefer his or her belief-system, but feels that he or she is compelled to subscribe to it because of their evaluation of the relevant scientific evidence that they are aware of.

It may be interesting to speculate for a moment what the psychological reasons might be for becoming a diehard atheist. Was it the painful slap on the back of the hand with a ruler by a nun in parochial school or an infatuation with an atheistic college professor or was there a "hanging out" with peers who consider themselves something like "the intellectual anti-religious elite"? And is his or her philosophical position often, occasionally, or rarely, a product of solitary, independent, systematic, and unbiased close reasoning? I am certainly not suggesting here that it never is, or couldn't have been!

Now, unlike reluctant atheism and agnosticism, diehard atheism is always reactionary because it is "anti-something." If one shares a dislike for the loud-mouthed preacher on TV – and I do, or has little tolerance for the pesky salvation religionist at work – and I have, it is easy to see why an atheist would occasionally want to "needle" members of the religious establishment a little when the opportunity presents itself. However, understandable or not, the reactionary attitude doesn't move us one stitch closer to understanding the Truth of ourselves and the Cosmos. "Fun" as it may be, it's still largely a waste of time.

So, is there any "non-belief" evidence for human survival after the disintegration of the biological entity?

The idea of reincarnation has been around for thousands of years. There are probably over a billion people on the planet today who believe in the doctrine in one form or another. The question of interest to us is: "Has any research been done in this area at a reliable and verifiable scientific level"? The answer is yes.

A pioneer researcher

Dr. Ian Stevenson, who around the nineteen-sixties and -seventies was a psychiatrist on staff at the University of Virginia, was a pioneer in the scientific study of reincarnation. One of his books, published by the University Press of Virginia, is entitled: *Twenty Cases Suggestive of Reincarnation* (ISBN: 0-8139-0872-8). After reading the book from cover to cover, one may well be left with the feeling that the best evidence yet for human survival beyond the grave is contained in its pages.

When we read about children of the Tlingit Indians of Southeastern Alaska for instance, who at a very young age begin to talk in great detail about how they were killed in a previous life as a result of bullet wounds, or by spear in a clan fight, and they can show the investigator matching-their-story birthmarks – entrance bullet wound pattern in the front of the chest with a "correct location" exit bullet wound pattern in the back, or a diamond shape spear wound pattern – and their stories are corroborated with painstakingly cautious methods (double-checking and cross-checking to guard against embellishment and fraud, etc.) in remote villages by informants who knew the personality of that former life, we know that we are dealing with powerful evidence of a certain type of human survival (pages 216-269).

What makes this evidence all the more credible is that Dr. Stevenson is clearly not a promoter of a belief-system. Even though he has collected, over decades, nothing less than a mountain of evidence that is – as he calls it – "suggestive of reincarnation," he has never been willing to say that the question of reincarnation has been settled. That mindset is the hallmark of a true scientist. On the other hand, he did say in an interview at one time that the collective evidence is strong enough to say that reincarnation, or "rebirth, as the Buddhists and the Hindus call it," offers us the most reasonable way to account for his scientific findings. Furthermore, unlike a lot of other scientific – legitimate or bogus – research in other fields, such as studies done to determine the effectiveness of certain medicinal drugs, Dr. Stevenson's research was never profit-driven. That fact in itself makes his findings less questionable because he was never beholden to any industry or shareholder. Of course, the other or negative side of that coin is that this lack of "bottom line" consideration, where nobody directly – at least not financially – benefits from any results, is, regrettably, the chief reason why so little of that type of work has ever been done.

A well-known actress talks about her former lives

So, why is Dr. Stevenson's work so different from the actress, Shirley Ma-cLaine – who not long ago traveled to Spain to try to "re-experience her former life as a Moorish girl in Medieval times" – telling us about her memories of her several previous incarnations, and from the former-life memory claim of General George Patton, commander of the Third Army in the Second World War, who "remembered" having been a (in-?)famous military leader "before" – just to mention a few?

Well, Dr. Stevenson's work is (was) based on veridicality and verifiability, while no other person – ancient or contemporary – that I am aware of, can make, or could have made, such a claim in reference to reincarnation.

Does that answer contain an implied criticism of Shirley MacLaine's announcements of alleged personal past-life memories? Most certainly not. However, inspiring as her personal revelations may be to her readers and audiences, her claims – in the absence of the possibility of authentication – would probably have been considered worthless from a scientific research standpoint. We must bear in mind that the human mind has a virtually boundless capacity to create fantasy formations – remember the "inventiveness" in your dreams – or to embellish a story that is otherwise essentially factual.

So, should we now say that we disbelieve the claims of the actress? No! We can leave it in the center – neither believing nor disbelieving them – because "there is no way to tell." On the other hand, an increasing awareness in society of this scientific research may reduce the general tendency to ridicule the person or the claims. It just may not be as "crazy" as we once thought. However, Shirley MacLaine doesn't appear to be overly sensitive – and, of course, she shouldn't be – to ridicule.

Various beliefs in reincarnation

Now, just because one accepts the reasonableness and the plausibility of the idea of reincarnation, doesn't mean that he or she has to abandon the "suspending judgment" frame of mind. It is still prudent to use phrases, such as: "If it occurs...." And above all, it certainly doesn't mean that we have to embrace the repugnant and devoid-of-all-common-sense versions of it, such as the Man-being-punished-into-the-incarnation-of-an-ant transmigration type of belief.

The "second chance" concept is an aspect of the reasonableness of the belief in reincarnation. It is consistent with the notion that if the individual "flunks" a semester or a whole year in school or the entire life cycle, one should go back and repeat the "course," or whatever it was that the individual should have learned the first time around. This benign and useful feature of this concept stands in stark contrast to the – not so appealing – belief that Man is chained, in the present life cycle as well as during many future incarnations before final liberation, to the wheel of Karma. Dr. Stevenson said in one of his interviews that, much to his surprise, he has never come across any evidence that supports the belief in the operation of a Karmic Law in all of his years of research. He added: "Of course, that doesn't mean that it does not exist; so far we haven't found any evidence for or against such a principle."

How do these beliefs "come about" anyway? Well, probably like this: first we have the cases (the testimony of the children or occasionally adults), and they in turn foster the basic belief in reincarnation. Then, over time, some of the believers begin to add to, or color the belief, thereby starting the process of manufacturing so-called truth which the cases themselves do not support. Then we have "a truth" or "a belief," which is not necessarily a reflection – much more likely a distortion – of "THE" Truth, or the Cosmic Reality of reincarnation – again – if it exists.

If it does exist, why doesn't everybody remember – at least in childhood – a previous life? That would be a little bit like asking: "Why doesn't every person that was ever born have the same level of musical talent"? Yes, we are all evolving but not necessarily in the same directions and not always at the same speed. But, maybe, this "remembering" is still more universal than we think it is.

A difficult to explain discomfort and a serial killer

If I may be so bold for a moment, I would like to share a personal apropos story with you. Since early childhood I have had an inexplicable reaction to a "stick shaped" object, such as a broom handle, being pointed in the direction of my eyes. When my wife mops the kitchen floor for instance, I feel that I "must" wait until she has finished mopping before I can walk through that area because the back and forth movement of the mop handle bothers me greatly. But my eyes have never even come close – my parents have also verified that with me – to being injured by any poking action since the day I was born. So how can one account for such a discomfort? Could it be a "worn down fear" resulting from an experience that happened in some type of pre-existence? Reincarnation furnishes us with a possible as well as plausible (not proven) explanation. It's certainly far from illogical!

Someone may ask the question: "Why don't we remember the events of a former life more vividly or in greater detail?" I, for one, am grateful that my past-life memory, if that is what it is, has faded to a very "vague level" or that it is "dim and out of focus." When we consider the incredibly inhumane acts of Man against Man from bygone times, who would want to have a "full-flavored taste" of being that victim – or even victimizer – again. Dr. Stevenson discovered that, far more often than not, the past-life memories are connected to intensely experienced, highly emotional events – such as a violent death, etc. Blessed forgetfulness? On the other hand, if someone with troublesome memories can be helped by past-life – even if reincarnation doesn't exist – regression therapy, the pragmatic recommendation can only be: "go for it."

It is interesting to wonder at this point – and of course this is nothing other than pure speculation – if the baffling and senseless phenomenon of the serial killer may be explained from the perspective of a previous life. I have lived in Spokane, Washington, for over sixteen years now, and during the first decade of that time, more than a dozen bodies of murdered prostitutes were found in the city, in the surrounding county, and in the Tacoma area. Robert Yates was arrested a few years ago and has since been convicted of these heinous crimes. There is nothing in his family background that can account for his behavior. By his own admission he had never been physically or emotionally abused. While incarcerated, a friend asked him at one time why he did it. His answer was a simple "I don't know." He is not the only serial killer that has answered that question in that manner.

The behavioral psychologist, who believes we all "started with a blank slate" and insists that we are what we are because of nothing more than the result of the total conditioning that took place since the day we were born, may have to eat crow here. The slate may not be blank at birth at all.

The "standby" brain disorder explanation would be quite a stretch too, because Mr. Yates, as a father-of-five family man and a helicopter pilot in the National Guard, was by all appearances a normal, fully functional member of society. Yes, a modern color brain scan shows an unusual pattern – predominantly blue rather than yellow, orange and red – in the frontal lobe of the brain in the serial killer, but are we looking at an EFFECT here or are we looking at a CAUSE? It doesn't make a lot of sense – it's far more likely that the killer's consciousness produces the altered brain pattern than the other way around – to identify that (the frontal lobe changes) as the primary cause! However, an extremely traumatic experience – such as intense suffering inflicted by a malicious and immoral woman (or women), or some other relevant calamity during a former life that resulted in a deeply-etched emotional scar in the psyche, which was subsequently carried along (like the Tlingit Indian cases, but this time at an unconscious level) into the next, or the serial killer's current incarnation – unquestionably furnishes us with the most reasonable, "fitting-like-a-glove" explanation. Of course it's pure speculation. I've already admitted that!

A British woman and a composer

A few years ago, in one of the investigative TV programs, such as "60 Minutes," "20/20," or "Dateline," an impressive case of reincarnation from Great Britain was featured. Unfortunately, I walked into the house too late to make a video-recording of that program. The case was about a woman in England who "remembered" a former life as a mother of four young children in Ireland. The most vivid part of that memory was that, in her former incarnation, she lay dying while struggling to try to hold on to dear life because of her intense or even paranoid worry (strong emotions) about the welfare and the future of her young children. She also had a detailed memory of a particular church façade in the town where she "once lived." In addition to the memories, she had a deep longing to be reunited with her previous-life children. Eventually she was able to travel to Ireland, find the town with that same church still standing, and meet "her much-older-than-she-was children," all four of them alive and well. Now, the children – being devout Catholics, who are "not supposed to believe" in that sort of heresy – came up with the "alternative" explanation that their mother in heaven had arranged this meeting with this loving younger woman, who knew all about their early years.

And how do we explain the child prodigy, Wolfgang Amadeus Mozart, who – as a virtuoso pianist since the incredibly early age of four and long before his feet could reach the piano pedals – gave concerts to the Empress Maria Therese of Austria as well as to audiences all over Europe, if not by supposing that there must have been extensive musical development "from somewhere" before 1756, the year of his birth?

(With my illustrations I can no longer hide the fact that I am a connoisseur of classical music. But, not to worry, I'm also a "kinda semi-normal fellow" with my liking of Elvis, Willie, Dolly and EmmyLou. Yes, some of "them fine tunes" are a delight to the soul.

Oh shucks, with these few lines I threw my last shot at a Pulitzer Prize right out the window. Oh well,…!)

A believer in reincarnation?

A reader may now be ready to comment: "So, then you consider yourself a reincarnationist, or a believer in reincarnation"? No, I do not! I neither believe it – NOR – disbelieve it; although I must admit that I now consider it more (perhaps

80% or so) likely than not that it does occur in some instances. Why can't we just look at all the evidence and say: "This is interesting and appears to be 'good stuff'; it's reasonable and it does make sense," and – then just leave it at that. Why "the rush to sign on"? Or, why be a believer of the flat earth theory in the beginning of the twentieth century when by the end of that century pictures from outer space have completely demolished that belief? And worse, why suffer from the extremely arrogant intellectual ignorance of the "supposedly omnisciently-wise and infallible top brass" of the Roman Catholic Church which condemned Galileo, forcing him to recant his Copernican view of our solar system. It only took about four hundred years for that fine institution to admit that the former leaders were wrong and to apologize belatedly for their awful treatment of the eminent Italian scientist. That very same closed-minded system also has firmly closed the door – probably because it contradicts the beloved heaven and hell philosophy – on the possibility of reincarnation. Suspending judgment until certain facts can irrefutably be established, has never been a part of their approach. Surely, we are all entitled to our opinions – but why not hold them lightly? Only proven science can establish infallibility, but, much to the chagrin of all the religious leaders around the world, religion – unless it is scientifically based – cannot.

On August 6, 2003, it was reported on the CBS evening news that a secret Vatican document was found, endorsed by the highest authority in the Catholic Church – which strongly suggests that it was signed by the Pope himself – that outlined the church policy on keeping the now well-known rampant sex abuse by pedophile priests a secret, and silencing the victims with the threat of excommunication. A lawyer, providing commentary on that CBS program, who definitely didn't create the impression of having "the smear campaign mentality," suggested that some type of criminality was involved in that policy. Wow – some infallibility! How does that rate on the evolution of consciousness scale? Oh well, the Vatican will "make a nice museum" when this is all over! Oh my …, here I go again, I got a little side-tracked.

In evaluating the reasonableness of the possibility of reincarnation, we may want to include the fact that certain baffling phenomena, such as homosexuality, cross-dressing, and the feeling of being trapped in a body of the wrong gender, become much easier to explain with the concept of "gender crossing" between sequential incarnations. Of course, that doesn't constitute a "proof." But may we put that down as: simply a good indication?

The soul-less reincarnation possibility of model 2

If model 2 (chapter 1) were the correct model of the Universe – the already discussed material strongly suggests that it is not – then a potential for a different type of reincarnation still exists. We may want to call it soul-less reincarnation. Unlike the above-mentioned kind of reincarnation, where we must assume the reality of a discarnate existence connecting the previous life of the individual with his "following re-appearance," there are no links (memories, character traits, etc.) whatsoever between the sequential lives (and "the possible simultaneous lives" [see chapter 2]) of the individual in this "soul-less reincarnation model."

How could this model even be possible? This is the reasoning: "Whatever happened once, for example, your own birth, can potentially happen again – because if it couldn't happen AGAIN, it couldn't have happened ONCE, but it DID happen once, because you are now holding this book. So, if you believe that pure Randomness (model 2) has put you on this earth, even this "unintelligent" Randomness can potentially do it again, because – here too – if it couldn't potentially do it again, it couldn't have done it once; and if this were true, that would mean that you do not exist – now. But you do exist.

You may object: "But the odds against this kind of reincarnation are too high." Under the law of probability, high odds against a certain occurrence are overcome

90

by an even higher "number of runs." If a lottery ticket has odds of, let's say, five million to one (against it being the winning ticket) and somebody like Bill Gates were to buy a billion (with a "b") tickets – which would be a "lousy" investment for him of course – it is virtually certain (the statistically probable number of winning tickets he would be holding is 200) that he has won (perhaps shared) the jackpot.

Let us assume for a moment that about 30 billion people on an average are born on our planet in a century (about five generations [20 years each]) over the next one hundred millenniums (science tells us that our favorite star – the sun – is "good for another five billion years"). That adds up to a total of 30 trillion (with "tr") people having been born after one hundred thousand years. Do you think that this is a sufficiently high number so that one of them could have "fallen together (randomness)" in the very same way you "fell together" in your current existence? That "one of them" is then "YOU reborn." So you may re-appear in the year 92,654. Of course, you wouldn't have experienced the lapse of time as a "long period of time," because non-consciousness (or non-existence) doesn't experience time or anything else, PERIOD!

And, of course, there is also the "preserved DNA revival" possibility, which would also be a form of "soul-less" reincarnation.

With a certain sense of amusement, we can now say that whoever has become aware of these two reincarnation possibilities, may no longer be able to speak the words: "What the heck, you *only* live once… (usually a "semi-apologetic" prelude to some kind of indulging)" with the same level of conviction as before, because even model 2 has its own reincarnation possibility – however meager (no accumulated wisdom) – attached to it.

A little reflection tells us that model 1's type of reincarnation, because of the obvious non-physical nature of a discarnate reality, adds some weight to its side of the scales. However, model 2's type of reincarnation – the soul-less kind – adds no weight to its side because it does nothing whatsoever to "increase the likelihood" of model 2 being the correct model of the Universe.

The near death experience

The mature and well-informed reader is probably aware of the large number of claimed so-called "near death" experiences. In recent years, extensive and well-documented hospital interview studies about this subject have been done, here in the United States, as well as in countries like England and the Netherlands. The results were the same everywhere. In many cases, the patient is pronounced "clinically dead" – the brain-wave monitor shows an absence of brain activity and the heart monitor shows the familiar "flat line (no oscillations whatsoever)" – and sometimes remains in that state for a considerable period of time until he or she is "brought back," or revived by whatever medical means. During that period of "zero brain waves," the individual may have one of the very common experiences of traveling through a tunnel towards an "unearthly" bright light, or may have an "out-of-body" experience where the "Self" may be looking down on its own physical vehicle lying on the hospital bed. Some patients have reported that in this state of "astral projection (a concept even known to the early Egyptians)," they were able to overhear the conversations of the doctors and nurses in the hospital room. After "re-entering" the body, the patients have often accurately commented on the contents of these conversations.

Now, if Man's consciousness is a product of the functioning brain (model 2, chapter1), how on earth could there be any kind of consciousness or experience at all when the brain (zero brain waves) isn't even "running"? Do we expect a stereo or a TV, whose power button is turned to the "off" position, or whose power cord is unplugged, to produce any kind of sound – however faint – through its speakers? The answer is so obvious that it doesn't need to be stated!

The diehard model 2 supporters might say: "To be "clinically" dead is not being "totally" dead, because in the latter case the individual couldn't have come back; and there may still be just enough residual brain activity – not detectable by a machine – in the former condition to account for these experiences in a non-meta-physical way (accusations of fraudulent research can no longer be made because the studies have been made by "too many reputable interviewing scientifically-minded doctors [and others] in too many different parts of the world," so that any "conspiracy theory" would become laughable and ridiculous)." The listener with-out an obsessive allegiance to a belief-system could then – perhaps sarcastically – ask the devout model 2 supporter if his or her explanation "might be suffering from a streak of a desperate groping at straws."

However, in our deliberation and evaluation, we should still pause and take that – however frail – "non-finality" objection seriously. It is true that the "near death" individual didn't die and that the body didn't disintegrate to the "beyond any repair" stage. That is why the approach of looking into many different direc-tions is taken in this writing, because in this way such a "deficiency" can be off-set by evidence from another area. The above-mentioned children of the Tlingit Indians of Southeastern Alaska, who were interviewed and examined by Dr. Ian Stevenson, certainly did leave their – left to disintegrate – bodies of former incar-nations behind permanently (pages 216-269).

Dr. Stevenson's work "backs up" and complements these reports of the near death experiences; and therefore we can say that it takes an incredible amount of intellectual stubbornness to continue to deny the foregoing some credence.

The often-quoted philosopher, Bertrand Russell, who didn't believe in the con-cept of survival after death, wrote in *Why I am Not a Christian* (first published in 1957): "For my part, I consider the evidence so far adduced by psychical research in favour of survival much weaker than the physiological evidence on the other side. But I fully admit that it might at any moment become stronger, and in that case it would be unscientific to disbelieve in survival" (page 45).

That was written half a century ago. Dr. Stevenson's reincarnation research was started a decade or so later; and much of the near death research was done much more recently than that. Has Bertrand Russell's "unscientific to disbelieve in survival" time now arrived? It's beginning to look that way!

The example of the automobile transmission may be helpful at this point. In all cars, whether stick-shift (manual transmission) or automatic, there are forward gears, there is a reverse gear, and there is a neutral position. If, in terms of the survival question, you're not quite ready to shift into a forward gear (to actively embrace the possibility of the continuity of life) are we, the better informed members of the human race, now not at the scientific turning point when it makes sense to – at least – shift *out* of the reverse gear *into* the neutral position. There are three positions and not just two. And by the way, there is no atheistic Inquisition than can force you to recant that common sense *neutral* position.

Two Presidents and a great yogi

If you still feel that you cannot possibly abandon the notion that physiological processes "pretty much have their way," and that the uncontrollable decaying process towards the end of one's life follows its own time-table, consider the deaths of our two early Presidents, John Adams and Thomas Jefferson. These two men, who became close friends at a later stage in life, "rigged" their time of departure from this world. They both "decided" to die – there were only a few hours between their respective transitions – on July 4, 1826, exactly 50 years to the day after the signing of the Declaration of Independence. Both men had a tremendous "emotional investment" in that event. Well, it is apparently not completely uncontrollable!

What is also significant is the fact that there is no evidence (from their correspondence) that they did anything physically to ensure that they passed away on that day and not on any other. They didn't take any preparatory steps, such as eliminating their daily vitamin intake at a precise moment – vitamin supplements didn't exist in their day – and they certainly didn't commit suicide by poisoning or starving themselves. Isn't that another indication that states of mind are primary and physiology only secondary (but by no means unimportant)?

One of the most amazing levels of control over Man's final departure from this planet was demonstrated by the Hindu yogi, Paramahansa Yogananda, who was a well-known spiritual leader in Los Angeles in the first half of the twentieth century. In the early nineteen-fifties, he announced that he would make his deliberate final yogic exit (or transition), called samasamadhi – giving details in advance, such as: "I will do this with my boots on immediately after addressing a large audience with a certain ambassador in it, etc. It happened exactly the way he planned it. According to the unanimous verdict of several medical doctors, his health was excellent at the time of his transition and there was no reason – no heart attack or anything else to die from – for his physiological system to have expired. He certainly wasn't "kicked out of life." Approximately one hundred thousand people paid their last respect to the great yogi and were witness to the fact that his not-treated-by-any-embalming-process-whatsoever body remained in a remarkably well-preserved, completely odorless state for three weeks before cremation.

The for-viewing-purposes open casket of another deceased man, who lived in a completely different world with a totally different mindset and philosophy, had to be closed and sealed after a mere three days because of the nasty stench in the viewing room. He was identified in life as pope Paul the Sixth. Interesting!

On the same page as the above quotation, Bertrand Russell speaks of the "terror thought of his own annihilation" (page 45). I seriously doubt that Yogananda entertained such a thought (and emotion) before his samasamadhi. As one lives, so one dies.

The reason that we can categorize Yogananda's extraordinarily well-documented samasamadhi as strong evidence for the continuity of life is that he is
the one fairly recent human being, that I am aware of (there may well have been
other less publicized samasamadhi transitions), who left this world from, or out of
a non-traumatized, non-diseased, and non-old-age-worn-out body. No "the brain
only" scientist can come even remotely close to giving a sensible explanation for
this astounding phenomenon. Consider the incontrovertible fact that if the physical
brain and the whole biological system are primary, and consciousness is only secondary or resultant, then a middle-aged man or woman in excellent health simply
CANNOT die or "get out," except by an unexpected medical failure (heart attack,
stroke, etc.) which definitely did not happen in Yogananda's case. The doctors
thoroughly examined his body – as any doctor knows, one doesn't miss the signs
of a heart attack – and found, as the yogi predicted, *nothing*. However, if consciousness is primary and what we are in essence, then it must have independent
existence (see chapter 2 for further elaboration) – non-contemplative people are
unsurprisingly, and even necessarily, unaware of that fact – and through development, such as meditative techniques and practices, Man can learn to exercise
control even over the "once considered impossible." This is all consistent with the
everywhere-in-life-observable evolution of consciousness principle.

So, the completely-unwilling-to-consider-a-new-idea-even-in-the-face-of-
overwhelming-evidence type of mind can only turn to another denial for comfort.
But it most assuredly *did* happen, because you cannot fool all of the one hundred
thousand within-a-few-feet eyewitnesses, many of whom personally knew that remarkable man. Even Hollywood couldn't have pulled that off. Remember: "You
can fool some of the people some of the time, … but you can't …."

In his writings, he described the path to take for this "process to get out." His
unusual development in consciousness clearly demonstrated possibilities that are

potentially available to every man or woman, because – again – there simply cannot be any special dispensation for anyone. Yes, even a single individual can prove an as-yet-unfamiliar (or not-widely-known) principle.

Devote yourself over a long period of time to the intense studying, practicing, and developing of just about any worthwhile goal in life – including this "life science" – and apparently the "ready" man or woman will rise into a noticeable level of mastery. The "stuck in the mud" individual, or the seemingly incurable life's-spectator-only type always looks pretty "darn" silly with his pronounced opinion of how far up the ladder of mastery one can climb. If you've never gone beyond standing on the sideline of life, you cannot possibly know what it's like to be an accomplished player. Or can you really?

Although I am not a yogi – and as a eclectic thinker I am not prepared to confine myself to any, however good, belief-system – I deeply appreciate Yogananda's contribution in showing us the kind of "stuff" and perhaps "non-stuff" that makes us what we essentially are. His assurances that there is a future life carries a lot more weight with me than any of the purely speculative pronouncements of the learned naysayers who are standing at an elevation level on the slopes of "Life's mountain" that isn't any higher than where you and I are. Why do I think that he must have been standing at a vantage-point, far above the crowds, where he could see what really lies ahead? Because of what he demonstrated and *not* because of any verbiage. There is wisdom in the expression: "Talk is cheap." Ralph Waldo Emerson said somewhere in his writings: "I've always appreciated people who can do things." I do too!

But then there is a question that rises up in the growing human mind which by its very nature shows up after – and rarely beforehand – these strong indications of immortality have been internalized. That question is: "Ok, fine, so the possibility that life goes on looks good, but is there anything we can do now to make that future life a more conscious one?"

Immortality is one thing but conscious immortality quite another. The former – if it exists – must be a principle in Nature, the latter – if it exists – would have to be earned.

According to the ones "in the know," there is something we can and must do. In order to draw closer and closer – it is an ongoing development and by no means an instant "one shot deal or affair" – to that ideal, we must, before anything else, learn to become more fully conscious in our present moments. The Now, or the Present Moment is all we have, and it is all that we will *ever* have. It is also all we need. When tomorrow arrives, it arrives as the Now and not as the future. The past and the future are images in the mind, but only the present moment is real and tangible. Is that common sense – or is that religion? Or both?

There is the old saying: "Don't just stand there – do something." No doubt, most of us are quite familiar – perhaps even too familiar – with that admonishment. Of course, we all know that there are times in life when immediate action is unde-niably required – we rightly expect the fireman and the policeman to perform as such during those critical times. But there are also times when physical non-action is – no doubt a completely alien concept to many – "the better way to go." There are times when it is sheer wisdom to reverse that saying to: "Don't just do some-thing – just stand (or sit) there." Those are the moments when the "inner reality of Man," which is the ordinarily silent, all-too-often-ignored essence in all of us, has an opportunity to become more active. That inner – consciousness expanding – activity is known by various names, such as: "Meditation (or prayer if you prefer the religious term), contemplation, practicing the silence, practicing the Presence, honoring the Sabbath – or simply "time out." However, that subject falls outside the scope of this book.

Oh, uh, one more thing

In the Western world, there is a deeply-rooted belief that the greatest tragedy that can befall Man is that he dies. And then there are those who feel that "not living right and not growing while we're here" is by far the greater tragedy. I am a "card-carrying member" of the latter group.

Like the one-raincoat-only, slightly cross-eyed TV detective, lieutenant Colombo, walking away a few steps and turning around with his fingers – that are holding his familiar stogie – on his forehead, saying: "Oh, uh, one more thing …," I feel compelled to mumble: "I was just wondering, sir, is it a good idea that we try to leave this world at a substantially higher evolved level than when we came in? … sorry, to have interrupted your train of thought, sir."

Chapter Six

The question of Statistics

This book is not a treatise on any particular or specialized subject, such as the principle of statistics, the complexity issue of DNA, reincarnation, or anything else that has been discussed so far. The primary purpose behind these discussions is to see which one of the two viable models in chapter 1 is supported – and to what degree – by these subjects.

Statistics are very "odd." Its application assigns a number or a ratio to a likelihood. In statistics we have a mathematical equivalent of quantitative words ranging from infinitesimally small to astronomically high. Only the mathematical or numeric side of the transposition (the translation to numbers) is by nature exact. However, because the side of the quantitative words is obviously inexact (it involves some rough guessing), the numeric side has a false or misleading exactness in the equation. So, if we have determined that a certain event is highly unlikely to occur, and we would then like to get an idea how likely it is that it will occur, let's say, three times in a short period of time, we can still transpose the likelihood to the numeric side, but we have to be – in the absence of an exact method of calculating the likelihood – very conservative in our estimation if we want to be credible.

The odds of a turtle egg

Let's apply this principle to an example. In the reptilian egg, five or more mutations had to occur simultaneously for the "whole system to work." For instance, if we watch – "live," or more likely on a Nature documentary video – a sea turtle lay its eggs in a hole on a beach, we know that the shell of the egg has to be strong enough to withstand the external pressure from the other eggs and the sand above it. It also has to be impermeable. After incubation the baby turtle has to have the tool, the egg tooth, and the pre-programmed (its parents are not there to teach it) know-how to break out of its limiting environment. There also has to be a mutation that allows it to absorb its shell food, and another totally unconnected mutation that keeps the waste separate from the nutrition, so that the developing turtle doesn't poison itself. So, here we have the five required simultaneous developments – which according to biological evolutionary wisdom can only arise from mutations: the condition of the shell, the egg tooth, the know-how, the ability to absorb the shell food, and the sealing off of its waste. If anyone of these five mutations would have failed to show up at that single right moment in evolution, or would have malfunctioned in the "collective (not in an individual one, of course) sea turtle egg," the species would have been wiped out.

Now we have to estimate the level of complexity – expressed in numbers – of each one of the mutations, and no one on this planet today can do that with any degree of accuracy. To be able to do that we would have to know the complexity of the "old series (DNA and the intricate structure of the particular complex proteins [see chapter 1] before modification – and we can only now begin to appreciate the awe-inspiring vastness of it all) of whatever it was that was modified," and the complexity of the "mutated series after the modification," so that we then have a detailed understanding (using a deduction process) of the complexity of the modification itself, which is what we need to transpose into a statistical ratio so that we end up with a mathematical likelihood number for our further calculations. Or, to put it into simpler terms with the use of a analogy: "we want to know, or calculate, how hard it was (or will be) to build an addition of a certain design to a particular existing house that we would like to remodel." So, we have to use a "ball park figure" estimate, and then we should reduce it to a very conservative level so we know we're not exaggerating the numbers; plain and simple!

The ratio of one million to one should be considered a conservatively estimated probability (against the appearance of a successful mutation) for each one of the turtle egg mutations. However, I would rather be ultra-conservative in this instance – so, let's reduce that ratio to one hundred thousand to one. The odds of each one of the mutations are then 1 (successful) to one hundred thousand (failures) or 1:100,000. Now for two successful mutations to show up simultaneously the odds then become 1:10,000,000,000 (10 billion). For five successfully INTERACTING – we don't just have five "independent" good ones, we have five "interdependent" good ones, which increases the odds dramatically – mutations, the odds are at least 1:10,000,000,000,000,000,000,000,000, or one in ten septillion (US; not British). And that's only the beginning, because there have been all kinds of simultaneous multiple interacting mutation developments in evolution whose combined odds may add hundreds and possibly even thousands of zeros to our last number.

Unbelievably improbable odds! I'm sorry, model 2 cheerleaders, but that's how the law of probability works.

If you, the reader, are still comfortable with these odds, you may want "to invest" in a few lottery tickets!? They all have far lower or more favorable odds than the reptilian egg.

So, why do otherwise highly intelligent people hold on to model 2 (chapter 1) when the just-mentioned odds are the square peg and Randomness is the round hole? I watched my grandson a few years ago trying to fit a square peg into a round hole, and when it didn't go in, he tried to pound it in with his fist. If he could have talked at that time, he might have said: "You little twit, get in….." But he was just a toddler. It's amazing that any grown-up would attempt to do the very same thing. Is it because they feel that they must go down fighting to defend their favorite model because it is considered dishonorable to capitulate? Interesting!

Now, if we consider these developments in the reptilian egg from the standpoint of model 2 (chapter 1) for a moment, then we may ask ourselves whatever happened to the 99,999 (the conservatively estimated statistically probable number) failed mutations of each separate development, anyone of which could have created havoc in the system. You see, for a single successful mutation to occur under the Randomness model, Randomness – if I may personify it for a moment – has to work overtime, while all the way along producing these thousands of miscellaneous "garbage mutations," before it can get to the one single good one, because that is its modus operandi. If that were actually so, wouldn't that be noticeable, and wouldn't the whole system have fallen apart long before the appearance of the one single "good guy"? So, it is not only the absence of the required fully functional mutation, but also the presence of the numerous bad mutations which Pure Chance would have had to produce (and aren't there) that could have annihilated these reptiles. As I have stated elsewhere in this writing, for unknown reasons bad or "disease causing" mutations do occur, but they are nevertheless relatively rare.

Another sound reason to refrain from calling these developments chance mutations. Only model 1 is capable of furnishing the system with the fully functional needed mutation.

The calculation may be exact but the interpretation is not

A software engineer may be asked to develop a program that produces random numbers with odds of one million to one. After completion, he really feels that he has created the best possible randomly-selecting computer program for his client. While he tests the software in his office, he sees the following numbers on the monitor: "381,468, 644,332, 879,212, 014,555, 000,517, etc. Hurrah, it's working! Then the client, who generally understands the law of probability, walks in and watches the demonstration: "091,113, 002,345, 084,460. At that point, the client's reaction to what he saw on the screen may well be: "That software is no good, because the selection process seems to be trapped within the one hundred thousand range"; while the software engineer says to himself: "Interesting, but that is still a normal run of random selection." Of course, when many more 6-digit numbers are produced on the screen that do not start with a zero, the client will eventually be satisfied that the program is working as it should.

The software engineer obviously had a "greater tolerance" for odd runs than his client. However, if he wouldn't have been at least a little nervous with the fourth number starting with a zero, we would not consider him to be an ordinary mortal. If the eighth number would still have been below one hundred thousand, the engineer would have been in *total* denial about his designed software. A visitor to a casino watching the dice showing the exact same numbers ten times in a row – and even long before then – would start looking for magnets, or some other clever device rigging the system.

So, the question is: "At what point does a probability run, or series, move from the "normal" range into the "abnormal" range, and, if continued in that direction, into the "impossible" range? The answer is: "It's all a matter of interpretation." Many people feel, as I do, that with the sea turtle egg mutations odds, the probability level – forever moving closer and closer to zero (expressed in fractions) but never reaching it – has sunk deep into the realm of total absurdity where we must begin to look for alternative " non chance" explanations, while others hold on to their beloved Randomness model with incredible tenacity, because they know that these wildly improbable odds are theoretically still possible. The question is not whether they are possible, because they clearly are, but are they reasonable? That all depends on whom you would ask!

By the way, doesn't that kind of tenacity remind you of the fierce Taliban warrior, drawing his sword, ready to do battle with an adversary using satellite-guided Cruise missiles and high flying bombers? He may have been brave and courageous, he may have been zealous and unrelenting, but he was something else too! No further comment.

Psychic phenomena

We must admit that the phony psychics with their advertised psychic hotlines on TV have dealt a serious blow to the level of respectability and credibility that the concept of psychic phenomena once had gained back in the days when Dr. Joseph Banks Rhine did his scientific research in extra sensory perception, etc., at North Carolina's Duke University. Regrettably, it is now largely considered the domain of the greedy and the unscrupulous at the one end, and the lonely and the gullible at the other.

The perceptive reader, undoubtedly, has observed the same trend in other areas: the "Real McCoy" unpretentiously enters the theater of life, and – in – walks the "wanna-be" a few moments later, who has dozens of brothers and sisters with a lot more hoopla and bravado. Oh, well, …!

However, "Truth" is not in the habit of "moving over" because certain groups of people tarnish the image of a certain aspect of it. There still remains much that we cannot account for by "normal" means and that "much" is not going away by ignoring it. For example, not a single human being has ever come up with a sensible explanation for the amazing capacity of a pet (cat or dog) to find its way – walking cross-country over distances of thousands of miles and periods of months without the assistance of a satellite guidance system – back to a former home or owner. So, the skeptic, unable – because of well-documented evidence – to deny it, just ignores it. Interesting and telling!

But before we consider the matter of the believability of the general concept of psychic phenomena, and the degree of veridicality of specific cases, we may want to examine our own attitude with which we approach this subject. Is there such a thing as "the inherently – or independent of anyone's opinion – best attitude" in this situation, and if so, what is it?

Let your opinions lie still

Once we have successfully peeled away all the nonsensical outer layers surrounding this controversial subject and have stripped it down to its core, we are then staring at the inescapable truth that the greatest credibility belongs to the determination not to "promote," not to have an agenda (for or against), and to an uncompromising resolve to steadfastly stick to the scientific posture of "suspending all judgment in the absence of an ironclad proof."

The Roman emperor, Marcus Aurelius, reminds us throughout his book called *Meditations*, that we should "let our opinions lie still, because we suffer from them as well as from our ignorance." Not very many people are good at that. And yet these insights of this remarkable man, who was – in addition to having been born to the purple – a great philosopher, are the most valuable pearls of practical wisdom shared with Mankind.

I sincerely applaud Mr. Randi, the well-known "professional debunker," and others, for exposing fraudulent claims and hoaxes. Not very many people today attend seances anymore since those sessions have been "disrobed" and "smoked out" as being *mostly* crafty performances of tricksters and charlatans. However, we must ask ourselves, once a hoax has been uncovered and the perpetrator has been unmasked and put on display for humiliation, wrists and neck locked in the pillory (figuratively, of course), what is it that has been proven (or proved)? Only one thing: "that some people 'do' hoaxes." Beyond that: Nothing! That's it! No more! Those of us who have "lived a few summers" should have arrived at the point where we realize that some people are "into that sort of thing," and since there isn't much that anyone can do about it – because it's not illegal – we are much better off if we do not over-react, because what was done may be "stupid," but isn't necessarily "outrageously evil." The "so-what?" attitude seems to be the best one to adopt here. It may keep the blood pressure down.

But there is still what appears to be "good stuff" – and I'm not prepared to throw out the baby with the bathwater.

Unconscious premonitions

In the Reader's Digest book called *Mysteries of the Unexplained* (ISBN 0-89577-146-2), an article was published entitled "Unconscious premonitions." This article first appeared in a Journal of the American Society for Psychical research. It makes reference to a study made in the nineteen-fifties by W.E.Cox who discovered that there are significant drops in the number of people traveling on trains on the day of a crash compared to the numbers on accident-free days. In one instance, there were only nine people on the train on that fateful day when there should have been between fifty and seventy (page 29).

Why does that study have some merit? The reliability of the "behavioral testimony" and the motivation of the individuals who had the premonitions cannot be questioned – precisely because they were unconscious. That only leaves the trustworthiness of Mr. Cox and his study. Now, if the man was "smart enough" to do this research, he was most likely also smart enough to realize that other researchers might follow in his footsteps, and either validate him with their similar findings or expose him as a fraud. Few researchers that I know of cherish the prospect of the latter. There is also no evidence that Mr. Cox was "on somebody's payroll." The only weakness in this instance may be the fact that this study wasn't extensive enough to make it a compelling case. But it is a good case and similar future research may eventually reinforce it.

This case – or actually cases – of the train wreck numbers indicates why the concept of psychic phenomena was incorporated in the chapter on statistics. In the process of evaluating the phenomena we are forever asking ourselves the question: "What is the likelihood – expressed in ratios – of the occurrences having happened as a result of "normal" or known principles, as opposed to them being a manifestation of "paranormal" or unknown principles. The calculation in this instance would be much more precise than in the later examples because we are already dealing with numbers which do not have to be estimated.

Is Nostradamus tumbling?

A most interesting debate was shown on one of the educational channels on TV. It was sponsored by Yale University and it centered on the prophecies of the sixteenth-century French physician, Michel de Notredame, more commonly known by the latinized name, Nostradamus. The program was called: "Nostradamus, a skeptical inquiry," which had a forum made up of believers and skeptics.

In the fifteen-fifties, Nostradamus wrote his predictions of the future in Old French in verses known as quatrains. Over the centuries his fame increased dramatically as a result of the claimed accuracy of his predictions of various events, including fairly recent ones.

So, who to believe: the believer or the skeptic?

One skeptic said that the predictions were more like postdictions. The language is so vague that it takes a tremendous amount of bending and twisting to make the text "match" the description of the prophesied event. An example of that twisting is that the name Hister is supposed to refer to Adolf Hitler. The explanation (excuse?) for the vagueness was that it was quite dangerous in that period of religious persecutions to write in a clear and straightforward manner.

Well, I agree with the skeptics, because this vagueness, with or without a – poor or valid – excuse, negates any claim to reliability and accuracy. It is like reading a horoscope in the paper where it makes little difference whether you read your own or the one of your friend who was born under a different sign. Yes, from time to time these columns in the paper may seem to hit the nail right on the head, but such an "every now and then event" fits quite nicely within the bounds of the law of probability (or the law of averages). Not too terribly impressive!

It's hardly ever a good idea to keep a belief alive by artificial means or by having to "prop it up" all the time, because if it can't stand up by itself, it may not be the thing that needs to stand up at all. Can you imagine having to defend the equation 2+2=4? It stands up no matter what you say about it, do to it, or how fiercely it's being attacked.

But maybe we shouldn't blame Nostradamus, because he may never have intended for "folks to pick up his verses and run off with them in wildly different directions."

Not to worry, Nostradamus may have tumbled – or more his supporters and devotees rather than the man himself – but the Cosmos isn't shaken by any such revelation, for it stands as firmly as our just-mentioned equation.

So, when a case has a weakness, we identify – unafraid and without any withholding – the weakness. When one perpetrates a "psychic fraud," we cheer on the debunker – the modern equivalent of the witch-hunter – in his pursuit of the scoundrel. But when we come across a good case, the strong features along with the possible remaining weakness(es) will be highlighted without exaggeration.

But here we do have cases of amazing accuracy

The following three good cases of extra sensory perception are published in the above-mentioned Reader's Digest book. Because I have not obtained permission from the Reader's Digest Publishers to use their material – which may not be easy to do since they, in turn, have procured their data from other sources who may still own the applicable original copyrights – I will only briefly describe the relevant-to-this-writing information in these three cases, which is permissible under the current copyright laws. (For those who wish to learn more details, I have already listed the ISBN number of the book in the earlier paragraphs with the heading "Unconscious Premonitions.")

Dorothy Allison saw the body of a five-year-old boy caught in a pipe in a vivid dream on December 3, 1967. She reported her dream to a police department in New Jersey. Three months later, on February 7, 1968, the body of five-year-old Michael Kurcsics was found who had drowned two hours *after* Dorothy's dream. Large pipes – in which the body had probably been wedged for a while – had been laid into the stream feeding the pond where the body was retrieved. All the details furnished to the police department by Dorothy – his shoes on the wrong feet, a nearby fence-encircled school with the number 8 on it, a gray house and a plant with a parking lot – proved to be 100% accurate (page 96).

Now, the usual complaint – and often a valid one – of the skeptic is that the information in a premonition isn't specific enough. The "self-crowned psychic," with the hefty per minute phone charges, will bring the "good news" of an upcoming promotion to the caller. That's easy to do since promotions are not uncommon events. But Dorothy Allison, who refuses (or refused) payment for her supplied information, was – to say the least – highly specific in her premonitions and her subsequent reporting of them. This unassuming housewife was by all appearances disinterested in promoting herself as a career psychic, and only sought to help for humanitarian reasons.

Although I am mindful of the fact that this event, which took place thirty five years ago, is extremely tragic – and Man is much more than a mere statistic – it is, nevertheless, legitimate (because it adds to our understanding) to apply the principle of statistics to this case.

How many five-year-olds walk out of their homes with their shoes on the wrong feet? Well, yes, it's not at all unusual that young kids make that mistake, but it is also quite common that a parent will catch that, or that a sibling will alert them to that fact, or that they – because of the discomfort – may notice it themselves when they start walking. Are the odds of 1:50 reasonable here? If not, lower them. The next odds decision is easy, because Dorothy saw a boy in her dream, and there are only two genders: the odds are 1:2. What are the odds of children *specifically* drowning in pipes within a certain time frame? Fortunately, these odds have to be considered quite high, and it's hard to believe that they would be lower than at least one in a one million per year for the whole of the United States, and one in one hundred million per year for that particular county in New Jersey; and Dorothy's precognitive dream and the boy's drowning were only two hours apart. Because Dorothy might have been dreaming about a middle-aged man, a teenage girl, the child having been kidnapped rather than having drowned, and an almost infinite number of "miss-matches," or not have dreamt at all, the odds of 1: 100,000,000 are conservative for that particular aspect (the drowning in a pipe) of the case.

Are you still pounding on that square peg?

If we decide on the reasonable odds of 1:1,000,000 for the fence-encircled school with a number 8 on it, and a gray house with a plant adjacent to a parking lot nearby, then we have combined odds of 1:10,000,000,000,000,000 (one in ten quadrillion). By comparison, all the (cash) money in the whole world today, expressed in U.S. dollars, doesn't come to more – I believe – than 100 trillion dollars, if even that. Ten quadrillion is one hundred times that amount. Truly staggering odds! It's a very strong case because of the specific elements.

Are there identifiable weaknesses in this drowning case? Yes, as always, the parties involved could have been lying about certain details, or could have made the whole thing up. But that would require a conspiracy of at least two people: Dorothy and a police officer (more likely a whole police department). She notified the police department immediately, but the boy's remains were not found until three months later. Dorothy's disposition and inclination (she has helped solve more cases) has already been alluded to, but why would a policeman or a police department want to "get in on such a deal"? Wouldn't it be far more pleasing to the men in blue to have such a case go down as having been solved by brilliant detective work? The thrust of their motive would definitely be in that direction.

True, the likes of Dorothy Allison are rare but not unheard of. There may well be dozens of perhaps well-intentioned psychics, who are utterly useless in solving cases for the police and society at large, for every Dorothy Allison. But we must remember that a weak case (or even many cases) doesn't *in any way* detract from a good case. The law of averages doesn't apply here, otherwise one strong case and one weak case would equal two mediocre cases. Nice try, skeptic!

It must have been my beef that pulled it down

During a bombing raid in World War Two, Winston Churchill got up from his chair at the dinner table where he was entertaining some government officials, and walked over to the kitchen where he told the kitchen staff to get down into the cellars immediately. He then quietly walked back to the diningroom to finish his dinner. Three minutes later the kitchen was completely destroyed as a result of a bomb explosion right behind the building.

On another occasion he suddenly decided to get into his chauffeured car on the opposite side of where he normally got in. Within a few minutes of driving through blacked-out London, a nearby bomb exploded, almost overturning the car. However, it righted itself a few moments later. Later on he said in his usual stoic manner: "It must have been my beef that pulled it down" (page 28).

Were the kitchen staff and the chauffeur "in" on the conspiracy? It could be,

112

but that's highly unlikely in such a high profile case!

Reasonable combined odds: "over one in twenty thousand maybe?" A good case but lacking the considerably higher number of details of the first example, hence the lower odds.

The crash of a DC-10

For ten nights in a row, David Booth had the same nightmare in which he saw an American Airlines DC-10 hurling (no longer flying) through the air invertedly and subsequently crashing. On May 22, 1979, four days before the event, he called the FAA, American Airlines, and a psychiatrist at the University of Cincinnati. Although David's information was taken seriously in some instances, there was little that anyone could do. On May 26, 1979, an American Airlines DC-10 crashed at the O'Hare Airport in Chicago, killing all two hundred and seventy five passengers on board. The crippled plane came in exactly as David Booth had predicted (page 32).

We know that we are dealing here with a definite detailed prediction and not a postdiction (see Nostradamus above) in this instance. Quite a few people – including public officials – in different organizations were notified four days in advance. That makes the conspiracy theory highly implausible. It is noteworthy that David Booth's dreams were vivid and persistent. Something was trying to get his attention, and it clearly did, although neither David nor anyone of us knows much about that "something."

Reasonable "ball park figure" odds: one in a million (not as many specifics as Dorothy Allison's case).

Combined odds

The combined odds for the three cases – and there are numerous well-documented cases in different publications around the world – are: 1:200,000,000,000,0

00,000,000,000,000 (1 in 200 septillion – without the numbers for the unconscious train wrecks premonitions). Am I anywhere near accurate? Of course not! There is obviously a lot of rough guessing involved. But the actual number is probably considerably higher. If you are not happy with my calculations use your own reasonable numbers; if you do it fairly, you will still come up with wildly improbable combined odds (remember to use multiplication and not addition!).

So, the evaluation process is analogous to what takes place in the quality control section of a factory where a product is checked for defects. If it passes the inspection it is placed with the good merchandise, and if it fails it is goes with the other rejects. Similarly, the above-mentioned phenomena are also closely looked at and judged on merit. However, I must admit that the approval process – where it must pass the "far more likely than not" mark – is, inevitably, much more subjective than its counterpart in the factory. After approval the case is tossed into the statistical pool for calculation purposes.

As was stated in the beginning of this chapter, the only reason for these exercises in numbers is to see which one of the two viable models of the Universe (see chapter 1) is supported by these statistics. The greater the odds, the more reasonable the assumption that model 1 is the correct model of the Universe. As I have said before, only the non-viable model 3 (chapter 1) is purely logical where the other two are not. It's "fun" to contemplate the reality of that model for a few minutes.

Merely a coincident?

The above-mentioned Reader's Digest book *Mysteries of the Unexplained* also contains numerous stories of remarkable coincidences. If they would be statistically rated, their combined odds would also be astronomically high. The over thirty coincidences connecting the lives of President John F. Kennedy and President Abraham Lincoln are both astounding and well-known; they are also exceedingly difficult "to explain away."

There are also many not-so-obvious, so-called coincidences in history. One of them involves the life and death circumstances of two famous World War Two generals, George Patton and Omar Bradley. Both men were effective war-time leaders, but had disparate philosophical attitudes towards the realty of military conflicts. Omar Bradley reputedly said to George Patton at one point: "George, you love this war effort too much!" There is good evidence that Patton "loved the battle" and thrived on, or reveled in the glory of his victories, while Bradley more or less looked at it as a "dirty job that had to be done." It was Patton who wanted to go on and "take on" the Russians, in order to wipe out communism. Thank goodness, nobody went along with him, because that "great Marxist experiment" nicely destroyed itself forty-four years later without millions of further war casualties.

George Patton died in 1945 immediately after the war as a result of a car accident, but Omar Bradley enjoyed many more post-war years and died not all that long ago when he was over ninety years old. Could it be that George Patton, with his "finest hours" behind him and no more glorious battles ahead of him, "subconsciously" decided that it was time to go, while the return to peace-time normalcy was welcomed by Omar Bradley? Who can deny that with any degree of certainty? And is there really any room left in the skeptic's bulging warehouse of denials?

An exhaustive analysis of a computation process

At this point I feel that I must warn my readers against reading the rest of this chapter unless they can determine for themselves that they possess the "type of mind" which enjoys an exhaustive analysis. The presented material may be too tedious, boring, and, as the word suggests, "exhausting" to any other "type." I have pondered for a long time on the wisdom of including such an analysis in this already-for-a-lot-of-wonderful-people-difficult-to-read book. Moreover, it is quite possible to grasp the gist (of the whole) of this book, even though a decision is made to skip reading the remaining paragraphs of this chapter. I wish that it were possible to make the analysis more entertaining (hopefully it still qualifies as being somewhat interesting), but I fail to see how that can be done. So, here you have – using the words of the Candid Camera TV show – "my fair warning" about the nature of the material "coming up."

A defense of the protracted argument

But first, I must say something in defense of the protracted argument (the non-quarrel kind!). Its virtue – if the argument is well-founded – lies in the penetration-deep-below-the-surface possibility, and obviously not inherently in its length. The desired depth is usually not reached until a series of relevant details are presented. One doesn't defend a thesis and earn a doctorate – if it's worth anything – with an extreme economy of words. In order to find a specific location or address, it's not only helpful, but often necessary to have – instead of a thumbnail sketch – a detailed map of the area or city. So, the supposed wisdom of "k.i.s.s. (Keep It Short and Simple – or Keep It Simple, Stupid)" – the expression was probably coined by someone with a very short attention span – doesn't always (not never) apply. The sound bite may have its place, but it seldom, if ever, leads to meaningful insight (perhaps, that is exactly what some politicians want). It usually doesn't do much more than either turn off the listener or confirm his prior opinions. The sound bite is a little bit like a brief stop-over-flight visit to a city on a longer journey by aircraft. In the early nineteen-eighties, I had to "hang around" the International Airport of Tokyo, Japan, for six hours during my return trip back to the States from India. So, if I were asked if I have ever been to Japan, I can truthfully answer that question with: "yes, I have." But have I "really" been to Japan? Of course not! I can tell you all about the airport souvenir shop and its bathrooms …. Many people visit arguments with only shallow stop-over-flight attitudes and souvenir-shop-and-bathroom insights. The useful pocket dictionary – handy as it may be – is *not* a "better" dictionary than the over two thousand pages Unabridged Webster; and that a detail-lacking brief argument is *always* superior to an in-depth protracted one … is a lovely concise bogus argument. There is a time and place for the shorter as well as the longer ….

So, in spite of the seductive practices by the merchant and the persuasion of the stimulate-the-economy advocate, I "k.i.s.s." my shopping list – and walk out of the store with nothing that wasn't on that list. However, your and my argument – providing it's "well thought out," important, relevant, original, stripped from trite expressions, and we don't ramble – should have a fair hearing by no-matter-what-size audience – and no "almighty" editor or literary critic will "get to k.i.s.s. that," or else he can kiss ….

While I'm writing this there is a war going on in Iraq. Predictably, such an event comes with anti-war protests in front of government buildings. One person (gender is irrelevant here) carrying a sign said to a reporter: "War is wrong, period – and there is never any justification for it."

Was the sign carrier right or wrong? Yes – and – yes! Over-simplification: a failure to recognize multi-faceted aspects and paradoxes in this serious issue – is the real problem here. Now, I will not be drawn into the argument about the justification for this particular war. However, it is generally accepted that there is such a thing as a "just war." Very few people in this part of the world would argue that the D-Day invasion in World War Two did anything other than serve a just cause. The depraved mindset of the leaders of the Nazi regime – there is little doubt that they sought the eventual annihilation of all non-Aryan, and even non-German people on the planet – made that awful, high-casualty event necessary.

What could be the reason that otherwise intelligent, morally well-developed people seem to be unable to acknowledge that military force is sometimes necessary, after all other unquestionably better and repeatedly attempted methods to resolve a conflict have ended in smoke? Could it be that they are the proverbial "children of light who are not nearly as wise in their generation as the children of the world"? Because that Bible passage (Luke 16:8) is so apropos and so extraordinarily well-fitting, I decided to overcome my general reluctance to quote from "the good book" and take proper advantage of the impact it may have on the living-exclusively-in-the-exalted-states-of-mind type of individual who has erroneously adopted the philosophy that all the issues of the world can adequately be addressed from within the walls of "the city on high."

But there is nothing in the universe that requires one to abandon common sense – ever. Yes, war is at once horribly evil, hellishly tragic, unbelievably stupid, but, nevertheless, sometimes – in spite of all our hopes and wishes to the contrary – unavoidable. Albert Einstein, who expressed the pacifist's viewpoint on several occasions, suggested at one time that if we (collective humanity) could – by way of wholesale refusal – limit military enlistment or drafting to a very small number (I believe it was 2% of the eligible men) in all the nations of the world, war would become impossible. Well, I – being a person who abhors violence – recognize the virtue and desirability of that idealistic proposition, but, beyond that, little practicality and – taking into account Man's present level of evolution – zero feasibility. Nature in its arcane wisdom has ordained that young men (and now women two), who – because of their youthful strength and faster reflexes have to do the fighting to protect the very young and the old as well as themselves – are generally (there are exceptions, of course) not as inclined to indulge in philosophical reflections – including mulling over Einstein's suggestion – as the older generation. That should be viewed – considering the dirty work they are sometimes asked to do for the rest of us – as their distinct advantage under combat conditions. Why?

If a loving mother decides to send her "Johnny off to war" with a heart filled with nothing but love and compassion for the enemy, and the army leadership had no opportunity to put more powerful alternative suggestions in the young man's head, he may hesitate and flinch at the crucial moment of taking aim and pulling the trigger. Paradoxically, a certain amount of momentary hate and contempt for the man in his rifle sights is his friend and protector. My brother, who experienced combat about half a century ago, told me afterwards that he had a hard time shooting at the enemy until he got "pissed" when one of his buddies got shot. So, the well-intentioned "gift" of the mother was made at the wrong time and under the wrong circumstances, and may possibly have cost the son's (or his buddy's) life. The "wisdom" of the head-in-the-clouds-without-her-feet-on-the-ground parent, which would have been so "right," wholesome, and congruous to try to impart *after* the soldier's return, was badly timed – and thus, in essence, sheer foolishness. Yes, without question, compassion and total harmlessness are indeed part and parcel of the "higher Man" – and the much-discussed evolution of consciousness process even demands an eventual movement in that direction. Still, the small-town elementary schoolteacher, who is required to teach in all grade levels, addresses his or her grade-six students in a different manner and style than the recent kindergarten graduates. Similarly, "life" must be appropriately addressed at different levels; and that awareness doesn't have to lead to any internal conflict whatsoever. If we know of a better way and can effectively (the knowing by itself isn't good enough) implement or incorporate it, then – by all means – let's not hesitate and use it. However – and this is the point where a lot of virtuous people stumble – if we cannot make it work, we have no choice but to go with an inferior option, when doing nothing is not an option. The unsophisticated "good vibes mom" didn't understand the rationale for that, and neither did the anti-war demonstrator. It's a manifestation of "incomplete seeing." Love is a great power indeed – but only wisdom may decide where it should and shouldn't go!

There is an evolutionary reason why it is still so unnatural for most human

119

beings to withhold judgment before "all the facts are in" and the last piece of the puzzle has been added. Like the well-known "fat storing gene," which was so beneficial to primitive lifestyles thousands of years ago – and has been such a nightmare to the "couch potato" in the last fifty or so – the rapid-assessment-and-split-second-decision-making ability also had a distinct survival advantage for our ancestors. To have investigated a perilous situation thoroughly, rather than quickly acting on protective hunches – which are prompting subconscious conclusions from the brain's memory database and not from a slow, curiosity-driven conscious evaluation of all the external facts – would certainly have led to their (and our) demise.

But Man is capable of both – the reading of the lines as well as the reading between them. His trouble lies in his native tendency to favor the latter over the former. A policeman can start an investigation on a hunch, but when the caught criminal is convicted, it was hopefully – regrettably, not always – done on the basis of detailed, well-scrutinized hard evidence; and oh how hard it is to overcome the pre-conceived notions and prejudices of the jury. Human opinions are often not much more than unchecked fantasies – formed by this unconscious we-must-quickly-fill-in-the-details evolutionary urge – "badly in need of some highly conscious updating."

If we do what the archeologist did, when he patiently brushed off and assembled the hundreds of pieces of a mosaic in anticipation of seeing the emerging faithful representation of what the artist of antiquity saw in front of him, we too may see a non-preconceived – and thus far truer – picture in any given situation before us, but *only* if we are prepared "to go the distance"; if we are – we may even discover some Cosmic artistry in it all.

Did I "go the distance" with my story of "the good vibes Mom"? Not so fast! There are other paradoxes (or twists and turns and seeming contradictions) in the larger tapestry of life that need to be considered. Even though it is unquestionably true that her unsophisticated approach to the reality of her son's combat circumstances is, humanly speaking, not "a smart way to deal with the situation at hand," there remains something highly constructive that she can do with "her good vibes," or "higher energy" states of consciousness. If it is true, that "Man is the supreme arbiter of his own destiny," and is potentially capable of making the Great Law of Attraction (see Chapter Nine for more details) work for Him – then that natural law is also available to her to use for the protection of her son. If she takes "the good vibes" and puts them, for the benefit of her son – time and time again, and even *tirelessly* – behind the unequivocal lines of the 91st Psalm (and you don't have to subscribe to any religion to claim the right to do this) that read: "A thousand shall fall at thy side, and ten thousand at thy right hand; but it *shall not* come nigh thee," until the conviction becomes unshakable – then there is an absolute certainty that her son will return home unharmed, because   unlike human laws – natural laws *cannot* be broken. If it is indeed a natural law – which means that its protective power is infinitely greater than any bullet-proof vest he may be wearing – then it seems like a good idea to become familiar with the workings of it – and "dwell in the secret place of the Most High" – well in advance of "a critical moment." If it depends on a certain religion, it may fail – if it's science, it cannot.

Am I suggesting that every argument should be drawn out (or sesquipedalian – teach your grandkids that one)? Of course not! Not every mosaic of life is worth looking at in detail. There are major and minor issues in life – and as one witty pundit stated it: "Too many people major in minor issues." With a minor one – brevity is always a virtue. With a major one, it may be wisdom to follow the thread (even if it's a longer one) all the way back to a certain point of origin while paying attention to what you're seeing along the way. There may be treasures at the end of that thread – and that thread is not a threat! If you suffer from a short attention span – take a few naps in between – picking up the thread where you left off each time, unless, of course, you don't care to know the end of the story.

Now, be honest, if I would have digressed miserly with a few scanty words, would the effect have been the same? I hope not.

It's time to return to our computation process

A few years ago, I heard of a statistician who claimed to have calculated the odds against our world and the universe having been created by chance. I do not know what methods he used and what numbers he came up with (they must have been astronomically high and couldn't have been more than very rough estimates). I have obviously already done some of that myself throughout the book, but I like to add a little bit to that by showing what all needs to be considered for such a computation.

On the very first page in chapter one, the only three conceivable models – with only two possible ones – of origin of Ultimate Reality are listed. Model 3 (the impossible one) earns by far the highest score of all three models in terms of likelihood of occurrence. If we imagine – anything being possible in the imagination – being there before that point of origin and had to logically predict which one of the three would prevail, number three – without question – would have been picked as the winner. That selection would, as a minimum, have eliminated the staggering intellectual difficulty of the concept of "beginninglessness." Now, as I have stated earlier, model 3 – notwithstanding being the logical one – is no more than an imaginable origin model. Absolute Nothingness never existed, because if it ever did – being non-transformable – it would have continued to exist (ironically non-exist is just as valid here); and that is obviously not the case since the universe is here and we are here. So, because model 3 receives the high score, we have no choice but to decide that the odds against model 2 (the Randomness model) are substantial right at that very early point of our computation process (model 1 [the Intelligence model] is not being considered here). Assigning an odds ratio to that early point is – considering our limited human intelligence – a difficult thing to do.

I will not try to identify everything that can be entered into that enormous computation process because that may well take more than a thousand pages of listing every known phenomenon or situation that would qualify. What I want to do here is: point to some of the eligible ones that are apt to be overlooked because of their familiarity – one of the great obstacles to alert observation and careful analysis.

Some time after the Big Bang, when – presumably hot – substance was hurled into the universe, stars of various sizes – our sun being one of them – were born. However – from the standpoint of pure randomness which we constantly have to go back to in this process – it makes no sense *whatsoever* to state that there was some kind of necessity that anyone of them would become a long-lasting (10 billion years – give or take …), life-supporting solar system star. No one can argue that they couldn't all have been massive stars – which, not unlike some of the Hollywood ones, burn up bright and relatively quickly (a million years or so) – or, the equally unsuitable Red Dwarfs, or Brown Dwarfs, or no stars at all (for example, huge clumps of ice, etc.), because the conditions and the properties of the various substances themselves might have been different. That they are what they are does *not* mean that they *had* to be what they are. That can be entered into our calculation. And then there are the planets which "didn't *have* to be there," unless …. And every last one of them might have shown up without that indispensable garment of physical life: our atmosphere; not exactly a minor detail. The list goes on – and on – and on. And, of course, all of that had to be in place long before the emergence of the first single cell life-form, which couldn't have materialized if the required various chemical affinities had never existed. That fact has an odds number. Again – did they *have* to exist? Familiarity answers "yes," but that habit of mind – I like to call it the lullaby faculty of the mind because it seems to put the rest of the faculties to sleep so easily – is sometimes wonderfully benign but frequently, I'm sorry to say, sheepishly stupid. Furthermore, it is "exceedingly and curiously odd" that everything that needed to work together – actually did so; and, of course, that holds true for all manifested life right up until the present moment. That last statement has nothing to do with religion as such – *only and exclusively* with an honest evaluation of numbers.

I remember the time, and maybe you do too, when researchers observed – probably in a laboratory – that certain molecules "accidentally" attached themselves to others to form more complex combinations of molecules that are basic building blocks of organic life. Not surprisingly, the conclusion surfaced here and there that that observation proves the Randomness theory and that all of life must have been formed by chance.

What are we to make of that? I do not deny in any way that they saw what they saw. What happened in their "soup" most likely did happen. But the conclusions drawn from that observation are flawed – because they are unnecessary. Why?

Picture, if you will, the following familiar scenario: we stand in front of a recently excavated site. There is a large pile of 2x4's neatly stacked on the left and to the right we see a stack of bags of concrete that are obviously placed there to start building the concrete basement. However, to the horror of the contractor, it had unexpectedly rained "cats and dogs" the night before. Water got inside the concrete bags, turning the 100 lbs bags into larger boulders. Aaaaaaaahhh, a random event in nature (the rain) transformed the powder substance into something that it was designed to become: hardened concrete. The observers in the laboratory merely witnessed the development of a synthetic process that apparently *readily and easily* happens in nature. That this process can be triggered by random forces (electrical discharges and what have you) should be no more surprising than the occurrence of the randomly activated event in our concrete bags. It is consistent with "the nature of the beast." What we – in my opinion – should wonder about is the presence of the basic elements in the "soup" and the odds of them being there or not being there; their absence would have been – by a country mile – more likely and logical. The lumber and the concrete are at the site for the purpose of building a house, are not there by accident, and very few people are puzzled by their presence. So, what are the carbon, hydrogen and oxygen molecules for? Why recognize the purpose in the building materials of the house and *not* in the ones of life? What the heck is the puzzling element in the latter??? The odds that the neatly stacked 2x4's and the concrete bags fell off a passing truck that, purely accidentally, took a sharp turn at high speed in front of that site, are – according to basic statistical facts – (although unbelievably high) millions of times lower than the accidental appearance of the scores of different building materials of life and the millions of other fine-working interacting factors. And this is only the reality consideration of all the stuff long before anything intricate and complex has been "assembled." And what are the thousands of natural laws, such as the highly useful laws of electricity, for? Just accidentally present? As John Stossel of ABC's 20/20 would say: "Give me a break." Do you need more reminders? I can – and I'm sure, so can you – easily fill up dozens of more pages, which I'm not about to do.

Because we fail to yet see an intelligence behind it all, doesn't mean it doesn't exist. After all, Mankind is only slowly "becoming more awake."

Since I have already given you my fair warning earlier, I would like to list in one "ugly (one of my favorite authors, the delightfully eloquent Ralph Waldo Emerson, would have shaken his head while looking at it) long sentence" an infinitesimally small percentage of these it's-far-more-likely-that-they-would-have-been-different phenomena, (natural) laws, and conditions of life and the universe(s). I'm not "trying to be cute" – I just hope that it confronts the reader with the enormity of this computation process. This is the one: It's far more likely than not that model

3 (chapter one) would have been the ruling model of Ultimate Reality, that at the time of the Big Bang only useless (unsuitable to support life) matter was thrown into the universe, that gravitational forces never existed, that huge asteroids would have wiped out all life during the evolutionary process, that diseases and other calamities might have done the very same thing, that all the needed biological functions never worked well (huge odds), that fuel to keep warm (let alone drive cars) was simply unavailable and that the stupidity of Man (not smart enough to understand pollution issues, etc.) would have made enjoyable activities – such as a nice walk along the beach – impossible for everybody.

That sentence has 638 characters or letters. Wouldn't you say that it's reasonable and conservative to think that an odds calculation (wildly inaccurate of course) of these few examples in that sentence has at least as many zeros in the ratio as the number of its characters? It is a pathetically conservative number if I had added more examples – which is very easy to do – that would have made the sentence two pages long. Now, using the 638 zeros number, the likelihood *against* the success (or *for* the failure) of these various events (combined) is roughly: one in one hundred trillion times a trillion.

I wasn't trying to create some abstract art with the visual appearance of the last sentence (although somebody may think it looks "cool"), but I *am* trying – with this long string of repeated words – to wake up the human mind (as much my own as anyone else's) to the outright absurdity of slothfully assuming that it's all randomness or chance.

I was amazed to learn recently how few people have a firm grasp on the concept of the number "a trillion" (a million million, or one with twelve zeros). No wonder so relatively few people are concerned about our national debt. The average man or woman in the street is asked: "What comes after a billion?" The all too common answer: "a lot – or a zillion – or something like that." Before you begin to chuckle at that, consider the fact that it is even more amazing that that odds number hasn't outright killed the randomness notion behind these examples. It should have, but it hasn't. Why?

An astrophysicist, who clearly reasons from a model 2 (chapter one) worldview, stated that he (unlike Carl Sagan, among others) considered it highly unlikely – in spite of the existence of countless billions (possibly even trillions) of other galaxies, each with their own billions of stars (it was recently estimated that there are ten times as many stars as there are grains of sand in the world) and the possibility of suitably attached planets – that there is extraterrestrial life anywhere in the universe simply because, according to his computation, the odds are astronomically high against that second (our world being the first …) occurrence. A most interesting conclusion – but there is a giant contradiction here that the proponent of that opinion may not be aware of. Our astrophysicist may well have had an odds number in mind that is not unlike my 638 zeros number. As a minimum, it must have, by far, exceeded the one in an octillion range – and probably many, many, many times that – because far lower ratios substantially increase the likelihood, under model 2, of there being extraterrestrial life somewhere, which he said he didn't believe in. But wait a minute, he knows that model 2 already produced one incredibly complex world under those astoundingly hostile odds – and, yes, they are way too high to have done it *even once*. If one lottery ticket buyer wins all the jackpots of all the lotteries in all the States that run them, and he or she does that three or four times in a row, nobody in his right mind would call that person lucky. He or she will be called either psychically gifted, or, more likely, an outright fraud. One of the two, but definitely not "lucky," because as we all know that the law of probability simply *cannot* produce that kind of luck. But the odds against the universe having been created by chance are – to say the least – every bit as high, just as far outside the range of reasonableness, and miles past the point where the unbiased individual would have, and should have (because logic dictates it), cut off the randomness explanation (I am well aware of the fact that I have followed the same line of reasoning with other odds calculations – but I consider it of such great importance that I decided to ignore the unwritten rules against the use of redundancy here).

So, why does the otherwise intelligent man or woman still hold on – frequently with astounding tenacity – to the randomness model? Could there be – since there aren't many other possibilities – some type of disorder, perhaps similar to the well-known obsessive-compulsive disorder (for example, people cleaning their doorknobs dozens of times in a day to guard against germs, etc.), which forbids access to the evaluation chamber of the mind where objectivity and rationality are also housed? O.C.D. is also known to occur in higher IQ men and women!

At any rate, these last two important qualities of the human mind are either missing or seriously impaired in the one who – after confrontation with the foregoing – continues in his refusal to postulate some type of – however finite or infinite – *a priori* intelligence (intentionally non-capitalized here) in this computation process.

P.S. Because one cannot write in a vacuum, a particular subject – which would serve well as an illustration of the protracted argument, in this instance: "the justification for the use of military force" – *had to be* selected. But the subject matter itself is obviously not all that important to the intent of this book.

Chapter Seven

The question of Mind Powers

The annual religious event of the Fire Walk has been observed and investigated by Western scientists. It takes place on Mbengga, one of the smaller Fiji's tropical islands in the Pacific Ocean. The participants prepare themselves for a considerable amount of time – prayers are offered to the water god, Tui Namoliwai – before they step out on the hot stones and charcoal. The pit is usually twenty-five feet long and about half as wide. The temperature is so high that the workers building the fire have to use long poles to poke it; the radiation heat is too intense to remain standing at the edge of the pit for more than a few seconds. If the walkers have prepared themselves well during the night, there are no injuries or even blisters on their feet after having walked – at a normal pace – the full length of the pit and back.

Over the years many theories have been advanced trying to explain how such an extraordinary feat is possible. In most instances an all-out attempt was made to try to "squeeze" the explanation into the framework of known natural laws. Of course, we should always first compare it to what we already know. However, in this instance, all of the "ordinary" explanations are unsatisfactory. At one time an investigator from the Smithsonian Institute suggested that, because stone or volcanic rock is a poor conductor of heat, no excessive amount of it is transferred through the skin of the "highly pain-tolerant" walker. That argument doesn't hold water. In the nineteen-forties, one presumably ill-prepared firewalker, who had to be pulled out of the pit by bystanders, burned his feet so severely that both of his legs had to be amputated to save the man's life. One man looses his legs and the one next to him walks the full fifty feet length without a blister!? Doesn't that largely eliminate the possibility of a "biological only" explanation? The two walkers, the successful one and the failure, were physiologically not *that* different! Besides, to prove that the pit is indeed intensely hot, a freshly slaughtered pig suspended on a chain had been lowered to a foot and a half distance – substantially higher than where the feet of the walkers were moving – above the stones and charcoal right in front of investigators on one occasion. The pork was served at dinner shortly thereafter. It certainly appears that the possibility of this event being a well-orchestrated hoax is virtually zero. Moreover, firewalks in one form or another – there are people who have walked barefoot on hot lava in Hawaii – are also successfully performed in other countries, such as Tahiti, Trinidad, Mauritius, Surinam, India and Japan. It has been done for ages.

A few years ago, someone told me about a firewalk demonstration – probably done in the nineteen-seventies – that was supposed to discredit the Fiji firewalkers (unfortunately, I have never come across any documentation on this). A similar size pit was dug and the charcoal was made to burn with the same reddish glow. However, the rocks or boulders that the walkers had to step on, were – unbeknown to the spectators – placed in the fire a short time before the demonstration began. How convenient! Unlike metal, rock needs a substantial amount of time to absorb the heat from its environment; a fact that I'm sure hadn't escaped notice within the group of "hell-bent-on-proving-them-wrong" organizers of the event. On Mbeng-ga, the fire is started the day before the walk and the rocks are in the pit right from the beginning.

I remember walking over concrete pavement a short distance from my motel-room to the swimming pool on a sunny summer day in Las Vegas one year when the mercury read 114 degrees Fahrenheit. On the way back to the room – being without any kind of footwear – I threw wet towels in front of me every few feet or so, to try to escape the extreme discomfort to my feet. By comparison, the burning charcoal reaches 1400 degrees and the surface temperature in the pit has been measured to exceed 400 degrees Fahrenheit.

So, who was doing the hoaxing at that "demonstration" and who was "trying to pull the wall over the spectators' eyes"?

What does it all mean? I really don't know, because I am just as dumbfounded as you are. However, one thing we know for certain: whatever it is, doesn't fit into a "neatly organized randomness-loving reductionist's package"! Yes, that much we do know!

This story from the Fiji islands and other places seems to suggest that Man can potentially acquire a state of mind which furnishes protection against extreme heat. Apparently, it is also possible to develop another similar – or perhaps even the same – state of mind which will allow the unprotected human body to withstand extremely cold temperatures for long periods of time.

The other end of the temperature scale

In 1981, Dr. Herbert Benson and a group of fellow scientists from Harvard Medical School were invited by the Dalai Lama of Tibet to watch, and investigate with their equipment such as temperature probes etc., Buddhist monks practicing their age-old "tumo meditation." In that meditative state, the monks can sit half-naked in the snow at sub-zero temperatures for hours. When a soaking-wet towel

– we all know what wearing wet clothes can do to the average man in these cold conditions – is wrapped around their neck and shoulders, steam will be seen rising from the towel after a few minutes. What was happening right in front of these medical men, wasn't supposed to happen according to orthodox science. Under even less severe circumstances, the survival mode of the physiological system in the ordinary human being quickly reduces the heat producing blood flow – first in the "non-essential" extremities, and subsequently in the "less-essential" muscles – in order to conserve heat, so that what is left of it remains available for the vital organs. The conditions that these monks voluntarily subjected, or subject (they still practice that today) themselves to, are simply not survivable for the average individual. As with the firewalkers, the notion that the monks, who were investigated by the Harvard team, may have perpetrated a hoax, is downright ludicrous. These oriental men now have an equally well-investigated kindred spirit in Europe who is validating their abilities in the twenty-first century.

A remarkable Dutchman

His name is Wim Hof. Like the Buddhist monks, this 42 year-old native of the Netherlands uses a yogic technique to regulate his body temperature. The Dutch firm, Doorgeest Koeltechniek, has built a special "exercise" freezer for Wim Hof in which he can sit, dressed only in a pair of shorts, practicing his yoga at the astonishing low temperature of –28 degrees Celsius or approximately –15 degrees (fifteen below) Fahrenheit. The sessions last about an hour and fifteen minutes at the present time. Fearing for his safety, the authorities (should they decide or should he decide?) have not allowed him to stay in longer. During these sessions, his physiological data are carefully monitored, minute by minute, with the most sophisticated equipment, such as infrared thermometers etc., by teams of scientists, medical doctors and film crews from the United States, France, Germany, Sweden and, of course, the Netherlands. He can stand covered to his neck in ice-cubes for

133

an hour without suffering any harm, such as hypothermia, frostbite, or anything else that ordinary mortals would experience under similar conditions.

In February 2002, Wim Hof ran in a marathon in Austria, barefoot and in his shorts, in sub-zero temperatures. While he was there, he also set a record swimming a distance of 60 meters (over 200 feet) under the ice without any insulating clothing. Then he decided, still in February, to bungee jump out of a balloon – with an unimaginable windchill factor – from a height of one kilometer, also in his shorts. One of his future goals is to climb Mount Everest – of course, smartly dressed in his usual attire for these occasions: barefoot and in a pair of shorts. His message: "We're not victims of circumstances and conditions; without the use of our native abilities, we only think we are."

Wim Hof is married to Caroline Hak and makes his home in Amsterdam. Apparently, you don't have to live in a monastery and adopt the celibate lifestyle to achieve such mastery over the human physiological system.

All of the foregoing, of course, flies straight in the face of conventional biological wisdom. But there it is! Unmistakably!

Swimming with the seals and the penguins

On February 12, 2003, the CBS investigative program "60 Minutes," featured a story about a swimmer with amazing but "curious (non-appealing – although inspiring – to most of us)" goals and accomplishments. Accompanied by warmly-dressed-and-safely-sitting-in-their-boats friends, doctors, the TV journalist Scott Pelley, and his camera crew, she (regrettably, I never caught her name) swam – in an ordinary bathing suit but otherwise bare-skinned – a half-hour, one-mile distance from a ship to the shore of Antarctica. The temperature of the ocean water was *below* 32 degrees Fahrenheit (salt water freezes at a lower temperature) and she was swimming between floating chunks of ice. What she did, so-called "cannot be done." She too is a medical curiosity, because death within five minutes caused by hypothermia is supposed to be a "certainty" for all human beings under these extreme circumstances. In her case there is no evidence that she uses a special yogic technique, but it is obvious that her states of mind of intense desire and incredible determination carried her to her goal.

What is it in us that can, if called upon, override or transcend the "normal"?

How can they do this?

So, how does this all relate to the theory of evolution? Should we be looking for a mutation again?

I can imagine if a professor of evolutionary biology would try to explain these extreme temperature conquering abilities with the favorite standby "just-another-mutation" answer, that one of his brighter students would scornfully whisper under his breath: "Yeah, right, give me a break! How far are you gonna carry this mutation thing?"

The polar bear and the arctic fox clearly derive their highly effective protection against extremely cold temperatures primarily from their fur adaptation, possibly resulting from a whole series of successive mutations since the time their distant ancestors lived in warmer climates. But, of course, here we are back in the realm of biological evolution or the "pre-evolution-of-consciousness stage", where adaptations are superimposed upon species and the (in this instance animal) individual plays no part – or at least certainly not a conscious part – in the development of them.

Some of the ancient thinkers have compared these stages of development with the principle of octaves. A purely universal or non-individual evolutionary system carries the creative process to the point of arrival of a well-developed brain, the lower stage or lower octave – then – this brain, or the mind, *must* be used to activate the individual creative process of the higher stage or higher octave in order to bring out its latent possibilities because the superimposing universal evolutionary process not only WILL NOT play, but by design CANNOT play on the keys of the higher octave. The higher octave is the exclusive domain of the individual. Good or bad, there is nothing in the Universe that forces Man to develop his innate talents or to play in the higher octave. One doesn't accidentally become a Wim Hof anymore than one accidentally becomes a concert pianist. True, there may be a bit of nudging from deeper levels of consciousness, but that nudging can be ignored, and that happens all too frequently.

What should become clear is that in the higher octave of the evolution of consciousness phase, it is no longer the advancement-causing mutation from the lower octave playing a transforming (or further advancing) part in that phase. Or to put it in simple terms: it was time for the mutation to retire at the beginning of the higher octave. Wim Hof's meditative technique clearly switches on, or activates, physiological processes, but not a mutation. Mutations are not de-activated and re-activated by the will of the individual.

It is important to note that these remarkable capacities were not "created" by the firewalkers, the monks, or Wim Hof. These individuals have learned to "use" what was already there.

Wim Hof didn't travel the relatively short distance from Amsterdam to Eindhoven to knock on the door of the research laboratory of Philips, the well-known electronic company of the Netherlands, to have them design and manufacture a special chip, to be implanted and hooked up in his brain by a brain surgeon, so that the humanly engineered electronics could control the re-activation of the various physiological processes that Nature, in its protective mode against the cold, had shut down. That would have been the man-made way. But if that were the case, we certainly would have heard about it. It couldn't have been kept a secret. Besides, the Buddhist monks have been practicing their tumo meditation centuries before our current age of advanced electronics and surgical implants.

One sometimes wonders if the average individual marvels more at such a human medical and technological advancement – if that ever could be done – than at Nature's astounding achievements. Maybe the old "familiarity breeds contempt" phenomenon will get in the way again.

The extreme pragmatist may say: "If I ever decide to climb Mount Everest, I'll wear a parka. So, I fail to see a useful application in the practice of that tumo stuff." Certainly, it is unlikely that droves of people will sign up to become members of "shorts only" mountaineering clubs (perhaps some of the members of the New Year's day "polar clubs" might be interested). But wouldn't this now well-demonstrated tumo technique – if mastered – be helpful to the millions of people who, for example, are suffering from poor blood circulation in the extremities? To remain free from the disastrous side effects of medicine alone seems to be worthwhile. So maybe it does have a practical application in other areas. But, above all, this powerful evidence indicating that we are far more capable than we once thought we were, has to be very good news to the evolution-of-consciousness-minded individual.

Lest someone interprets this expressed exuberance as evidence of a conviction that the abilities of the remarkable human beings mentioned in this chapter are unlimited, a comment is necessary. I am well aware of the fact that the firewalkers do not rest their feet in the red-hot charcoal, that Wim Hof does not sit in his freezer for days (at least not at the present time) at minus 100 degrees Fahrenheit, and that the female swimmer wasn't in the freezing cold water for hours. However, any unbiased and fair-minded man or woman must admit – unless, of course, they refuse to look through the figurative microscope – that the respective incredibly hot and extremely cold resistant abilities of these highly focused men and women are extraordinary.

Do I think that these phenomena are mystical or miraculous? No! I simply believe that we are dealing here with laws and principles that we are presently unfamiliar with. That's all! It's certainly not "spooky."

However, again, it is difficult to see how these manifestations can be reconciled with the Randomness model of the Universe.

P.S. No other so-called mind powers (premonitions have already been dealt with in the chapter on statistics) were included in this chapter because the evidence for them was, in my opinion, not strong enough to warrant such inclusion.

Chapter Eight

The question of Rejection of Religion

Because the central theme of this writing is about finding – in an unbiased way that truly stands up to logical analysis – and highlighting evidence pointing to either: a) the operation of Pure Chance, or: b) the existence of an *a priori* Cosmic Intelligence, and because rejections of religious ideas, or religion itself, may lead to an unnecessary and unwarranted denial of the latter (choice b), it is useful to show that there are viable alternatives to these rejections. The perceptive reader will have noticed that the use of the name, or, more appropriately, the word "God" has been avoided in all instances except in quotations. Unlike the expression "*a priori* Intelligence," the word "God" has many connotations and may conjure up different images according to one's belief-system. One may think immediately of the God of Abraham, Isaac, and Jacob – or Allah – or Brahma, Vishnu, and Shiva – or, less likely, Manito. As was stated in earlier chapters, no religious belief-system whatsoever – valid to the believer as it may be – will be promoted in these pages. Everything that is being discussed must stay within the parameters of recognizable universal principles and laws, and must measure up to common sense reasoning; and whatever doesn't meet that standard won't be discussed. Ideas of a speculative nature that are set forth here and there in this writing are deemed to fit within these parameters precisely because they are presented as speculative and not as anything doctrinal. Doctrines often come and go, but unadulterated common sense reasoning usually has a lot of staying power.

Relating honestly to purposelessness and unpredictability

If my evaluation of the evidence for or against the existence of an *a priori* Intelligence is as unbiased as I intended it to be, I must spend a moment with a paragraph or two relating to the phenomena of purposelessness and unpredictability. The well-read, scientifically aware reader is probably somewhat familiar with the Heisenberg principle. In a nutshell, it says: "the whereabouts of a particular particle in a gas cannot be predicted, and is nothing more than a statistical probability," i.o.w., we can never be sure where that particle "will hang out" at any moment in a volume of gas. Einstein – at least initially – found that a disturbing discovery (he probably got over it after a while), which – according to some accounts – made him exclaim: "Der Herr Gott doesn't play with the dice!"

In order to decide – fairly – if the "whereabouts unpredictability phenomenon" must be viewed as a strike against model 1 (chapter 1), we must ask ourselves the following pragmatic question: "Even though it is a legitimate scientific pursuit to try to learn why we are looking at "chaos" instead of "order" in a volume of gas – shouldn't the question of usefulness of the particle predictability be considered above all?" Would it make "Life" function better if it were predictable? If the molecules in the twenty dollar bill that I need to hand over for purchases in the grocery store would behave like the ones in a gaseous condition, then the slip of paper with the portrait of Old Hickory may disintegrate before it reaches the hand of the cashier. Now, I don't know about you, but I would find that very, very disturbing – not only because that would make "Life" malfunction, but also, because

I have never been able to persuade the Secret Service to allow me to print these not-always-easy-to-come-by, highly useful slips in my own basement. But, seriously, who cares where a particular – since they are all identical anyway – oxygen molecule was hiding out on the day before it was sucked into the intake manifold of your or my car engine, or before it entered the bloodstream to sustain your or my life. Heisenberg adding weight to the side of model 2? Forget it!

A popular American author and the Almighty

Samuel Langhorne Clemens, better known by the pen-name Mark Twain, enjoyed great financial success in the second half of the nineteenth century as a world-renowned writer, storyteller, humorist and satirist. The by-all-appearances happy Clemens family lived in a large lavishly furnished home, aided by their seven servants, in Hartford, Connecticut, during the eighteen-seventies and -eighties when Mark Twain wrote his most popular books about the to-most-of-us familiar characters, Tom Sawyer, Huckleberry Finn, the Prince and the Pauper, etc.

After the death of his twenty-four year old daughter during his absence, and increasingly so after the death of his beloved wife, Olivia Langdon, Mark Twain turned in thought against his Christian God, in addition to blaming himself for the loss of his family members. He also had a hard time reconciling the notion of a Providence of Love and Benevolence with the hypocrisy, racism, the hunger and poverty, particularly of children, and other misfortunes that he saw all around him. Even in his bitterness, his wit and humor were still present when he wrote in later years: "The Almighty and I are no longer on speaking terms."

Yes, how does one reconcile suffering and misery with the belief in Goodness, Mercy and Love?

When that question was directed to the average Christian theologian in Mark Twain's days or even in more recent times, it usually met with the following not-too-terribly-informative and not-to-be-questioned-any-further pet answer: "Well, you know, the Lord God works in mysterious ways." I am not suggesting that the man of the cloth wasn't sincere, or that he was arrogantly dodging the question, because that may well be the only available, and therefore best possible answer to give if one reasons from the belief in a personal, anthropomorphic Deity. However, it is obvious – or even superfluous to say – that such an answer doesn't do much for the millions of people who, without being as vocal about it as Mark Twain, for the most part lost interest in religion.

The evolution of consciousness and the octave

In the last chapter the analogous-to-the-stages-of-evolution "octave" principle was introduced. The concept is important enough to elaborate on it some more.

A musical instrument, such as a piano, has several seven-note-series or octaves. Hitting any key in a lower octave sounds the same as hitting the eighth key to the right, except the latter has a higher pitch than the former. If we compare the whole biological evolutionary process to a lower octave, then the following octave – immediately to the right on the piano – may be compared to the evolution of consciousness process. As was explained in the last chapter, the individual plays no part in the forward movement or progress during the biological evolutionary process, and thus doesn't play on, or strike the keys of the lower octave – but he or she becomes the principal player of the higher octave.

Of course any analogy has its limitation. The situations, qualities, conditions, or circumstances that are being compared in an analogy may be similar but are never identical. In the biological evolutionary process it is clearly observable that the "winners" in natural selection are the species of increasingly higher levels of intelligence. In ascending the musical scale – moving your fingers to the right on a keyboard – the pitch or audio frequency becomes higher; hence the parallel. In a higher octave a new series begins at the precise point where the lower one has just ended. A melody played in a lower octave, starting with the C note for example, can also be played in a higher octave starting with the C note, located seven keys to the right (the eighth key) of the one of the lower octave. So, like the first key, the eighth key is the beginning of a new series.

In the biological evolutionary process or the lower octave, the universal powers of initiative and selection are manifested as natural selection. Similarly, the individual has the prerogative of exercising his powers of initiative and selection in the higher octave of the evolution of consciousness process. Individual selection begins where universal selection – not in the Cosmos of course, but in the personal arena of the individual – leaves off or stops.

In the personal evolution of consciousness process, if it is to take place at all, the individual must begin by – using the most suitable imagery – putting on the mantel of his or her own sovereignty. Of course, that sovereignty or rulership is over one's own consciousness only, and not over the minds and lives of other people. Before the mantel can be put on however, there must first be a recognition that this internal act of the mind is an absolute and unavoidable prerequisite to any further progress.

In order to better understand the foregoing, we must realize that in the biological or universal evolutionary process all the forward developments are first and foremost for the benefit of the TYPE, or the particular species, and that at that stage the welfare of the individual members, or the single life, is of relatively little consequence. In the human species, generic Man is important at that stage and even zealously defended where the individual man, woman, or child clearly is not.

So careful of the type it seems

So careless of the single life

Does that mean that all the world's teachings are wrong in stressing the importance of the individual? Most certainly not! The key to understanding this paradox is the realization that the well-being of the single life is – only – relatively unimportant in the lower octave. Not until his or her arrival in the higher octave, or the evolution of consciousness stage, is the individual raised onto the plateau of preeminent importance, and it is right here that his further personal evolution – including the forward thrust towards higher and higher levels of individual attainment and perfection with its corresponding gains in personal joy and happiness – becomes the supreme focus. However, perhaps much to the chagrin of the many who are accustomed to – and may even be comfortable with – drifting on the tides of life, this higher octave is very much a *non-automatic* personal initiative stage.

There is no evidence whatsoever that there is anything in the Cosmos favoring Wim Hof (see chapter 7), as some sort of "teacher's pet" or "favorite son" of a Cosmic Intelligence, and as such exclusively awarding him – and thereby unfairly bypassing billions of others – his unusual powers and control over his physiological system because he was such a nice and deserving man. The only reason that he and the Buddhist monks acquired their uncommon protection against extremely cold temperatures – whether or not that ability impresses, or appeals to you is an entirely different matter – is that they practice something that you and I don't. That principle of personal "stick-to-it" practicing must be (or have been) incorporated by anyone who has ever demonstrated any great skill or extraordinary ability in any direction; and we will never know – with the possibility of previous lives – how long their practice run was before they got to that point. Is it still more likely to you, the reader, that talent is the sole product of a "lucky" arrangement in the genetic code rather than being the result of previous dedicated practice in the direction of that particular talent, done somewhere, considering we now live in post-Dr. Stevenson's-research-times (see chapter 5)? The other way around perhaps? In any event, there is no special dispensation for anyone, and no one can point to any evidence that any individual in history ever had a special status or unique kinship with the Cosmic Intelligence, which was not available to all other men and women.

The higher octave player creates his very own special status – the lower octave player "does no such thing"!

To explain this concept of concern for the type and lack of concern for the individual in the lower octave a little more, the analogy of the automobile manufacturing plant may be effective.

In the highly competitive field of automobile manufacturing, it is the demanding task of the designers and mechanical engineers to build the-best-value-for-the-money car models that have few, if any, design flaws so that financially disastrous recalls may be avoided. If, during the development period, a certain prototype does show a design flaw, the developers go back to their drawing boards and redesign the prototype – or possibly design an entirely new one – which is then tested under the performance conditions. At that stage, there are no cars with serial numbers coming off the assembly line that were built after that particular prototype. However, once that car model goes into production, a few of the first units are extensively road-tested for performance and reliability before production goes into high gear.

Now, it is obvious that the developers are extremely concerned about the performance of the car model that they designed. However, whether or not a particular single car with a certain serial number has a unique mechanical problem, possibly as a result of the fact that somebody wasn't paying attention at the assembly line, is of very little concern to the developer, because he knows that that situation was entirely outside his area of responsibility, and that there are auto-technicians in a different department of the plant that can correct these problems. So, the developers or designing engineers are exclusively concerned with the model or type, and the auto-technician takes care of a possible mechanical failure in an individual car. If a car owner has a mechanical problem which does not originate from a design flaw in the model of his car, he shouldn't direct his complaint to the car designer, but should contact an auto-technician instead. However, if the car owner insists to speak to the car designer anyway, he or she would be correct in answering the complaint with: "Sir (or Madam), I'm sorry, but you are barking up the wrong tree."

Similarly, if we assume the correctness of model 1 (chapter 1) for a moment, then we can say that the Cosmic Intelligence, which is the driving force behind the whole biological evolutionary process, doesn't concern itself directly with any-thing other than preserving the species – whether animal or human – but has nev-ertheless provided for the human individual in an indirect way by supplying him or her with a highly developed brain and a corresponding intelligence, which include the powers of initiative and selection, with which one can carve out one's own destiny. Hence the expression: "The Cosmic Mind cannot act FOR (individual) Man but It can only act THROUGH him."

Was Mark Twain – and so many others with him – "barking up the wrong tree"?

Isn't it more useful to search for the causes of failures within the compound of the consciousness of the individual rather than outside of it, now we have been introduced to the logic that individual failures are not universal or design fail-ures? The Designer cannot use the universal powers of initiative and selection to fix individual human problems, only the local technician (the individual himself), using his own powers of initiative and selection – albeit, as always, with the aid of universal principles – can. We just looked at the logic behind another familiar expression: "God can only help those who will help themselves."

If the greater or happier human experience can only come from a better un-derstanding of certain creative principles which can only serve Man if he or she draws them inside his or her own personal compound of consciousness, then the focus should be on that internalization process and it should be seen as a fine waste of time and energy to – continuing the imagery – bark at anything that is outside the fence of that compound. And even if Man had no access to the just-mentioned principles – which is hard to imagine in a Universe of seemingly infinite possibilities (and there are many, many people, past and present, who have disproved that notion) – then the old idea of "resignation" is still superior to a continued "kicking" or complaining, because that will *certainly* get us nowhere; although some people give the impression that useless complaining is a "fun process in itself."

Although it is true that this unavoidable or non-bypassable transitional phase – from the lower octave to the higher one – is a difficult one, that makes it all the more important to travel through that passage as quickly as possible. However, it is not the purpose of this book to supply "personal development recipes, such as 'the ten (or whatever number) steps to … program'" or to make any recommendations, such as taking a yoga class for example, because I trust that the reader, now armed with the knowledge that there is such a reality as the evolution of consciousness process, can find his or her own entrance to the path to be walked on. But some walking must be done if unnecessary suffering is to be avoided.

Do the religious wars disprove model 1 (chapter 1)?

There is an argument that some members of the atheistic community – undoubtedly because they believe it carries a lot of weight – frequently pull out of their collection of evidences against the existence of a Cosmic Intelligence. We could call it the religious wars argument. The rationale behind that argument – in a nutshell – can be stated as follows: "Some of the most horrible wars have been fought, and some of the most inhumane acts of Man against Man have been committed in the name of virtually all the major religions. Because religious teachings are supposed to make Man 'good,' and that, in general, has happened so far, religion is ineffective or even dangerous, and therefore God does not exist."

On July 7, 2001, an article by Shankar Vedantam about so-called "spiritual experiences" was published in Spokane's largest newspaper, the Spokesman Review. It was called "Scanning for Spirituality." One of the brain researchers mentioned in the article, Michael Persinger, a professor of neuroscience at Laurantian University in Sudbury, Ontario, Canada, was quoted as saying: "Religion is a property of the brain, only of the brain and has little to do with what's out there." The article also suggests that Persinger is one of "those who believe the new science disproves the existence of God and that they are holding up a mirror to society about the destructive power of religion, because religious wars, fanaticism and intolerance spring from dogmatic beliefs that particular gods and faiths are unique, rather than facets of universal brain chemistry."

Well, it may sound all so rational, but a little reflection upon the reality of that uncompromising statement by Persinger shows that his attempted leap, from the brain to "what's out there" (or God), didn't really get him to the other side of the gape (gaping hole), or across a wide gap in logic.

An example may demonstrate the size of that gape. Let's imagine that there are a lot of people around the world who believe that the spiritual center for Man on this planet is located in a large divine stone, buried a thousand feet beneath the icy surface of Antarctica, exactly at the South Pole. Since nobody actually believes that, nobody will be offended by this example. Then there comes a time when it is possible to send a scientific expedition with the proper equipment to get that stone out of the ice. Of course, then there are the two possibilities of finding a stone or rock, or of not finding anything because there was nothing but ice. Let's assume that a stone was found, which was subsequently taken to a laboratory for testing. The stone is placed in a cage with laboratory mice with miscellaneous diseases to see if the stone could effect any kind of healing. Nothing happens. It's just an ordinary stone. What is it then that we have proved (or proven) or disproved? What conclusions may legitimately – in the sense of what is logically permissible – be drawn from that? Only *one* thing, and *one* thing only: that that particular belief was a figment of the human imagination. Nothing whatsoever beyond that! The Universe doesn't stand or fall by or with the grace of Man's beliefs. The principle of pure Logic – which is just as real as the principle of mathematics – doesn't permit Man to make a connection between the now discredited or destroyed belief in the special divinity of a stone and the existence or non-existence of a Cosmic Intelligence. The gape is simply too wide to jump across.

When the Russians sent their first Sputnik into orbit and nothing resembling the "Man upstairs" image was seen in outer space, they thought they had confirmation of their atheistic worldview. Again, they only didn't see the "Man upstairs." That's all! The image that they presumably were looking for in Space – but to their delight did not see – of a large Michelangelo's "gray-haired, bearded God the Father" type of figure, was the image from the belief – which is totally inconsistent with the core teachings of their own sacred scriptures – of the majority of Christians and Jews (not the Buddhists, and others).

Now, if there is – and everything, so far, is pointing in that direction – an *a priori* Cosmic Intelligence, It cannot possibly have, or be limited by any extensions – as the Creator of these two realities – in Space and Time. So there cannot be a "physical or spatial image" anywhere. Moreover, the notion of a "God living in a spatially higher place, or Heaven, above the earth" stems from the wrong understanding of the meaning of the word "higher." It is a word with multiple meanings. It can signify a degree of a reality in addition to the spatial meaning. The word "higher" in a sentence, such as: "Man has a higher level of intelligence than a fish," is an example of a non-spatial, but very legitimate and correct meaning and use of the word. And, of course, a Cosmic Intelligence must necessarily be infinitely greater or "higher" than Man's intelligence, hence the use of the word "higher." When the weather forecaster talks about "higher temperatures," everybody knows that he is not talking about temperatures high above the earth. But when children are asked where they think their great grandmother "hangs out," the majority, even in this day and age, will point a finger – controlled by a religiously conditioned brain – to the ceiling. And astoundingly, unbelievably, amazingly, incredibly, astonishingly that "spatial heaven notion" lingers on – and on – and on.

Not long ago, on the Opinion page of a local newspaper, an atheist made the philosophically highly immature demand to be shown a picture of God. In a responding article, I asked the gentleman if he could send me a picture of Gravity, not what it does or how it works – and thus a picture of an apple falling out of a tree wouldn't have qualified – but what it (intrinsically) IS. I got no response. I am not wondering why. I know why!

No beliefs or belief-system, however wholesome (having a kernel of truth in it), primitive, or even "wacky," does anything other than show what Man is capable of fabricating in his head. The debunked divine stone story, as well as anything else that has ever been debunked, doesn't add any weight (as is so often assumed) to either side of the "likelihood of the correctness of Reality models" scales (see chapter 1). No belief-system, or the debunking or dismantling thereof, can do that, PERIOD! Reality is simply not subject to Man's beliefs and doesn't bow down to his imaginings; and therefore the disproving of a belief doesn't have one iota to do with what is ultimately true and what isn't.

Persinger believed that, because he was able to "trigger so-called spiritual experiences" with the use of magnets around a subject's head – and thereby bypassing the meditative techniques that supposedly lead to the experience of the "larger Self" – he had arrived at the "ultimate proof of the non-existence of God." These frequently classified as "religious or spiritual trips" in the head, whether induced or spontaneously occurring, whether real or imagined, whether useful and meaningful or not, have nothing whatsoever to do with the proof of a Reality model.

Again, these so-called spiritual phenomena do not prove or disprove the existence of a Cosmic Intelligence or Supreme Being anymore than that faulty mathematical calculations affirm or deny the principle of mathematics. We do not throw out algebra and trigonometry – I'm sure some "math-hating" students think we should – just because we've come across something that didn't compute. A mistake never denies a principle. And I don't even know if we are dealing with mistakes here, because I don't have a clue of what all of that "head stuff" essentially is and what it represents. I don't think many others do either.

There is only one "full-proof" way – that I am aware of – to get to that "ultimate proof of the existence of an *a priori* Intelligence." It is done by trying to gather enough reliable evidence from different directions which accumulatively and progressively shows that Randomness (the only alternative to *a priori* Intelligence), with its "utter non-intelligence" as well as its exceedingly strong tendency to return things to chaos, couldn't have "pulled off" the creation of the astounding complexity of Life that we are – in this book as well as everywhere else in life – looking at. Only when that has been done successfully, and not until then, will we have – through the process of elimination – arrived at our "ultimate proof", because – again – there are absolutely no viable alternatives to the first two Reality models of chapter 1.

Do we already have enough evidence? Calculate the odds against Randomness (see chapter 6) and print out the number and then decide. But stop by at your local stationary store first and get an extra printer cartridge and at least two reams of paper. You're "gonna need them" to get that odds number on paper!

Furthermore, if Randomness or Pure Chance, sitting behind the steering wheel of the creative process ever since the time of the Big Bang, could have had nothing more than the mere level of ordinary human intelligence at its disposal (of course, it would then no longer be Randomness), could It – even then – have "pulled it off"?! Who – among the ones capable of objectively evaluating the intricacies of life – would be so bold as to answer that question in the affirmative?

Being acutely aware of the fact that people often have a fast and indiscriminate reaction to what they hear or read, I must elaborate a little bit on the earlier use of the words "permit" and "permissible." The reader who, at an earlier time, has been introduced to the basic principles of logic – with its strict rules on drawing (or abstaining from …) conclusions, will know what I'm talking about. When I say, my grandmother has a sweet and round face and that's why the earth is round, it is obviously legally permissible to say that; thank Goodness for the First Amendment. It is also socially and ethically – and in every other way – permissible to say that, but, it is logically *not* permissible to say that. No connection can logically be made between the shape of my grandmother's head and the shape of the earth because no logical connection exists between these two; the gape is too wide to jump over or across. The words were used in the same sense that it is mathematically not permissible to say that two and two equals five.

An experiment in the imagination

It would be wonderful and enlightening if a certain experiment could be conducted which would settle this "religious wars" argument once and for all. Unfortunately, this experiment cannot be made in actuality, but we can speculate about the outcome of it in the imagination.

Going back to the time of the early Egyptians, we inject all human beings on the planet with a special substance that exclusively takes out all their religious ideas and sentiments. So, from that moment on, not a single person can think about gods and devils or heaven and hell any longer. The substance also has the characteristic of passing on its effect to all future generations. Then, in modern times, we carefully study the world history of these five millenniums or so. Would we then still be looking at wars and cruelty here and there in the various centuries, or would we be looking at paradise-like conditions everywhere? Well, we know for certain that there wouldn't have been a Roman Catholic instituted Spanish Inquisition. But the "lovely" tools that they used, such as the infamous rack and the hot pokers and all the other monstrous torture devices we have on display in museums around the world, would they still have been invented as the awful means to punish – now not in the name of religion of course – and control the unfortunate?

Surely, because it is an imaginary experiment I can't prove the outcome of it one way or the other. But who would answer this torture tools question in the negative? And then the wars question? And wouldn't there have been cruelty just the same, except we would have to designate it as "non-religious cruelty"? Would that designation have made the slightest difference to the victim?

The two incredibly tragic World Wars of the twentieth century were primarily territorial wars. Before and during World War Two, the Germans, or more specifically the Nazis, were looking for more "lebensraum (room to live)" for themselves and dominance for "das Herrenvolk (nation of superior men – as they saw themselves)" and racial purity of the Aryan Man – and religion was not really much of a factor. Their hatred of the Jews was based far more on their notion of the Jew's racial inferiority than on the various rabbinical teachings as such. Thank Goodness, the collective consciousness of the post-war German people no longer includes an appetite for war. Evolutionary progress? Yes! Still at a snail pace? For now, yes!

The fanatically devout Catholic British Queen, Mary the First, known as Bloody Mary, had the unbeliever burned at the stake during her short reign. However, her father, Henry the Eighth, who went down in history as an extremely despotic and cruel man – remember his two unfortunate Queens, the second and the fifth wife, going to the scaffold in the Tower of London, and the other two, the fourth and the sixth wife, cleverly but narrowly avoiding that same fate – was only conveniently religious and not particularly devout at all.

So, in light of the foregoing, doesn't it make far more sense to blame these horribly inhumane acts on Man's primitive, or poorly developed consciousness – or being near the low end of the evolution of consciousness scale – at a certain stage, rather than on his particular religion, or religion in general. Religion as such has so little to do with it.

A careful study of the history of the subject reveals a general decline of repugnant practices and attitudes, such as cruelty, hatred, intolerance and the appetite for war, in much of the Western world, especially in recent times. There is a pronounced reluctance in the civilized world to send men (and now women too) into combat – and it is certainly no longer done for the pure pleasures of honor and glory. The primarily secular human rights philosophy has also found a firm foothold in the democratic nations of the world. However, according to recent polls and surveys taken here in the United States, the level of religiosity of the general population – no correlation to church attendance – is still, and consistently remains fairly high. When asked if religion had a significant meaning in their lives, well over eighty percent of the participants answered "yes" on some of these questionnaires. Surely, some people answered in the affirmative out of a sense of guilt for not living up to their own standard in this area, but it is, nevertheless, safe to say that the average man or woman in our part of the world is not – at their core – a non-religious or irreligious person; and that there is only a small minority of atheists and agnostics.

Now, if it were true that – as Persinger asserts – religion or religiosity inherently fosters or promotes cruelty, bigotry, intolerance, aggression and whatever brood of negatives we may think of, why isn't there a corresponding decline in religiosity when there is clearly a substantial and measurable general decline in cruelty and all the rest of it? In other words, if a direct cause and effect link existed between religiosity and cruelty, then a decline in the latter should certainly produce an equal decline in the former. That hasn't happened in the past and is *not* happening now. However, a link between cruelty and a lack of moral development (what is, agreeing with Bertrand Russell, by no means an exclusively "religious" or "religion produced" quality or virtue) or "humane enlightenment," clearly and obviously does exist. Yes, in the days of Henry the Eighth, almost one hundred percent of his subjects declared themselves to be religious, but we all know what happened to you back then if you didn't. That's like the one hundred percent vote that Saddam Hussein got in the last election (2002) in Iraq. "Go figure!" It's unlikely that the average individual in the sixteenth century, regardless of his or her religious external practices or observances, had a greater internal affinity with religion than the average modern individual. But cruelty was rampant a few hundred years ago, while – at least in our Hemisphere – it clearly is not now!

Persinger and the others unquestionably pinned it all on the wrong cause. Even though their argument may initially and superficially have looked good, it is the step by step analysis that – as it frequently does – revealed the flaws. Again, if we must nail it down, we can say, that it is the old raw Neanderthal character, or Man's remaining – and in some individuals still dominant – reptilian part of the brain that gets in the way rather than anything having to do with religion as such.

It isn't all that long ago that the intense studies of chimpanzee (our distant ancestor) behavior, done by the well-known Jane Goodall, revealed that senseless violence predates Man and thus predates religion by millions of years. One day she – much to her horror – observed a gang of male chimpanzees coming down the side of a mountain, chasing a group of smaller chimpanzees – females and juveniles, and when they caught up to them, assaulted the weaker ones in an unbelievably violent and cruel way. This display of seemingly purposeless (from an evolutionary standpoint) behavior – which obviously far exceeded the punishment-for-the-purpose-of-correcting-delinquency level – shattered Jane Goodall's belief in the basic harmony of the pre-Man species. It should also shatter any man's notion that cruelty and religion are inextricably linked, because, as far as we can tell, the ape has no religion. Violence and cruelty exist – no more with religion than without it.

"Well, you are quite a defender of religion," somebody might say. I most certainly am not! I most assuredly do not consider myself a friend – but by no means an enemy or some sort of antagonistic activist – of organized religion in general, and orthodoxy, doctrines or dogmas in particular. And even though I would rather be in the company of an atheist or an agnostic (they usually reason more intelligently) than in the presence of a pushy religionist, I am neither an atheist nor an agnostic. But when a wrong cause is identified, one must speak up. And I did!

Are there absolute values or is it all relative?

In the second half of the twentieth century, and most notably in its free-love-and-anything-goes sixties decade, the suggestion gained grounds in some circles that there are no absolute universal values and standards. It became the age of the discovery of the "Self," and many, mostly in their teens or twenties, saw much of life as being purely relative to the fancies of the individual. "Good" became "Bad" and "Bad" became "Good" – and it was all the same.

Now, there are historians who are, or claim to be capable of identifying cause and effect relationships between culture phenomena of various eras. I am not one of them. However, that doesn't disqualify me from saying that this trend, to some degree, may well have been a reaction to some of the stifling orthodox religious attitudes – remember the forbidding of dancing in some organizations – of pre-World-War-Two times. Yes, the pendulum lingered for a while in the period of dominance of the extremely conservative orthodox religious mind-set, which required one to struggle against all kinds of condemned natural impulses (including the quite harmless ones), but its return swing in the direction of casting off these burdens and greater personal freedom was inevitable and predictable – requiring no crystal ball for the man or woman of insight.

One of the virtues that "got a bad wrap" somewhere along the line was Love, or, more specifically, Altruism. If one is "heavily into doing one's own thing," and has – perhaps unconsciously – adopted the survival-of-the-fittest or "dog eat dog" attitude and worldview, then it is perfectly logical to become an Ebeneezer Scrooge (before his "conversion") towards others. It is logical and understandable because religious preaching frequently aimed at trying to persuade the follower to practice the highly unnatural – and therefore largely unappealing and thus "unstable" – "laying down of one's own personal interest" from time to time in order to be helpful to, or mindful of others. The religious sacrifice notion dictated that you must give up something, or lose momentarily, so that others may gain. But the conflicting natural tendency says: "Helping an old lady cross the street is one thing, but paying the hired help a generous wage reduces my own financial level of comfort, and I may even end up being destitute myself." Self interest and altruistic interest were seen as direct opposites and thus mutually exclusive. Because it was "difficult to live up to," one had to internally "screw oneself up trying to be good," or "whip the ego" in order to put this religiously required virtue into practice. No wonder this "saintly" concept of Love has not (universally) ranked high in popularity so far. It is therefore easy to see why some folks decided to strip Love of its status of being an absolute – reducing it to the anything-goes relative – value.

However, if Love is a veritable absolute universal value or standard, and has archetypal existence, then its reality or status *cannot be modified* by human fiat. It can be banished out of the human mind, but, if it has archetypal existence – it cannot be banished out of Reality.

For the benefit of the reader who is unfamiliar with the concept of the "archetype," I must elaborate. From time to time one comes across writings in which the author states that archetypes are either irrelevant or do not exist. That can easily be disproved in a few sentences.

The concept of the immaterial archetype behind any physical thing or phenomenon was introduced into the body of philosophy by Plato. How can we prove Plato's premise? We could use the simple example of the reality of the circle. Suppose a silly law was passed, demanding the destruction of everything that is round. The coins in our wallets become square, we brake all the dinner plates, and soccer must be played with the oval-shaped football, etc. What have we accomplished then? We have destroyed and annihilated all "the circles – notice the presence of the definite article 'the,'" however, we haven't in any way affected, or done anything to "Circle – notice the absence of the definite (the) and indefinite (a) articles." We have only "wiped out" all the manifestations of roundness, but (the archetype of) Circle was completely unaffected by it. At any time we choose to, we can pickup two sticks, tie a string between them, and – voila – we can draw a perfect circle again in the sand on the beach. Yes, the "copies" can be destroyed but the essential "Thing itself" cannot! There is nothing whatsoever that humans can do to eliminate the latter. Hence the expression: "It cannot be touched by human hands."

As far as we can tell most animals are incapable of entertaining abstract ideas, but most human beings – although perhaps not accustomed to it – are nonetheless capable of it. Applying it may be a good idea.

Have we now not irrefutably proven or demonstrated the existence of a particular archetype, and thereby the reality of The Archetype. It's not particularly mystical, is it?

So, how do we go about proving that Love is an absolute or universal (independent of the human mind) value, virtue, standard or quality? In a similar simple way in which we can prove that 2+2=4 is an absolute mathematical truth. The basic proof is in the fact that *it works* and 2+2=5 doesn't work. Our calculations come out "right" with 2+2=4, where 2+2=5 jams the whole "caboodle."

Similarly, Love works. Not because it is "nice" or "sweet" or "the goody-two-shoes," or even having anything directly to do with the concept of "doing the right thing," but rather because its use or application consistently gets "positive" or "life affirming" results and it never jams anything.

An illustration is required. Of course Love encompasses many areas of life. I want to confine myself to "the love of one's work." I recognize that "love in relationships" is at least as important, but that subject requires more space than I am prepared to give it.

Not long ago, the ABC investigative program, "20/20," aired the unusual story of a privately-held specialty software company. Regrettably, I do not remember the name of the – I believe California based – organization. This fairly large company provides some unique amenities to its employees. It has its own daycare, staffed medical clinic and dentist office, a complete gym and other recreational facilities, psychologists and family counselors on the payroll, motel type guest rooms, a car repair shop, and a host of other pleasantries all there to make the employee comfortable.

Now, why would any owner/CEO in his right mind "burn up" that much of the company's profits just to be – altruistically – extraordinarily "nice" to its workers? That is a logical first reaction question, which – as it turns out – is, nevertheless, based on a false assumption and an erroneous assessment. The owner of that company and man of uncommon insight and vision had correctly determined at some point that this investment in his employees is easily offset by a general boost in productivity with less required supervision, an increase in loyalty towards the company, a decrease in absenteeism – mostly due to far less sickness (real and faked), and a substantial reduction in the very expensive process of hiring and training new workers.

How do we know that he was correct? His company has done quite well over the years, and even remained profitable during general economic downturns when his "not like-minded" competitors were hurting or even going out of business. The owner's double-edged sword type of altruism and "enlightened self-interest" were merged in his vision, and these two values, which – as I have stated earlier – are generally considered to be complete opposites, became one and indistinguishable, because it is impossible to tell from which one of the two he reasoned and acted.

Love works and thus the "higher" life works. And this is pure science because it is demonstrable, repeatable, and utterly reliable, and it has nothing whatsoever to do with persuasion or preaching. Religion seeks to promote the virtue of Love, but it is science – in this instance, the mathematically-exact predictability of a working principle – that proves its universal value. Who wouldn't want to "get in" on an employment deal like that? The list of people wanting to work for that software company and its "saintly" owner/CEO is about a mile long; and they will be on that list for a long time, because – unsurprisingly – nobody is quitting.

This is only one example of course, however, research has shown that whoever loves his or her work for whatever reason, "has a far better shot" at staying healthy, being more energetic, receiving timely promotions, avoiding pink slips during a general layoff (management will usually do their "darnest" to hold on to that man or woman), and enjoying a better homelife – and, by the way, there is no "nasty flip side to this coin."

I wonder what would have happened if the preacher throughout the ages would have presented *that* picture to his audience instead of the customary sacrificial-struggling-trying-to-be-good one? Might there have been more "takers"?

And how this story stands in stark contrast to what was happening in Birma during World War Two. The Japanese captors, responsible for building a railroad in that country and in a race against time, were trying to squeeze some more pro-ductivity out of their exhausted POW slave labor by "cranking up" – what they were good at – their level of brutality. The marginal increase at the expense of their prisoners was easily offset by – what they probably hadn't counted on – an increase in incidents of sabotage. I suppose one becomes more courageous when faced with the inevitable prospect of dying under the most miserable of circum-stances as a mere "consumable" of Imperial barbarians. With that awareness and little hope of escape, the dangerous decision to try to throw a monkey wrench into the machinery of the enemy naturally became much easier.

Apparently, the supposedly all powerful – expected to override everything – survival instinct in the human psyche didn't accomplish in that instance what Love readily and easily did in our specialty software company story. Moreover, the performance of other crude, "grinding" motivating forces, such as guilt and hard discipline, is equally dismal compared to the performance of free-of-internal-and-external-friction-and-thus-not-slowing-down-anything Love (or simply "Inner good feeling"). Should we really be surprised that a "higher" method out-performs a "lower" one? Wake up extreme disciplinarian!

That's why the – for some people hard to say – four-letter "L word" stands for an archetypal absolute Universal Value and Principle, and not because I (or anyone else) said so. I wish I could take credit for the creation of that highly effective "L-thing," but it was already patented eons ago ....

In these last few paragraphs, this virtue proved Itself as an immensely effective universal value, which sooner or later must be incorporated into the consciousness of anyone who wishes to become "the higher edition of himself or herself." All I did is describe how It did it.

It's not relative!

Practical religion can protect you against road rage and being "flipped off"

A few years ago, two friends of one of my sons were traveling on the freeway within the city limits of Spokane. Because they were engrossed in conversation and minding their own business, the driver hadn't noticed immediately that the aggressive driver in the car in the center lane was trying to overtake them and cut them off in order to get to a shortly-coming-up off-ramp. A very familiar scenario these days. Because my son's friend hadn't reacted within the exceedingly short "time-frame expectation" of the aggressive driver, the man "had no choice" but to show the middle finger to my son's friend. However, the latter – to protect "his honor" and not to be outdone – "had no choice" but to flip off the provoker in return. That gesture sealed the fate of the back window of his car. The two friends subsequently noticed that the enraged man was following them, so they tried to "lose" him by going around a few blocks a couple of times. That didn't work. It almost seemed like the bumpers of the two cars were welded together. So they then decided to pull up to a roadside Espresso stand, I guess – silently hoping that the man didn't care for cappuccinos. He pulled up right behind them, got out of the car with a tire-wrench in his hand and smashed their back window, while yelling: "If you got the balls, you get out of the (bleep) car …." Fortunately, one of the other Espresso patrons with a cellphone had quickly called 911, and the I'll-teach-you-a-lesson man, who was so "deeply insulted and emotionally distraught" by the lack of respect of his fellowman, was handcuffed a few minutes later and enjoyed the rest of his day in the back of a patrolcar followed by the registration and pic-ture-taking session at the desk of the busy, high-security-with-free-meals, one-star (almost) downtown "hotel."

What does this story have to do with religion? Well, not a "darn thing," if you're thinking about altars, church steeples, sanctuaries, the heavenly choir –

priests, nuns, and rabbis – and "sinners in the hands of an angry God." But – it has *everything* to do with "the condition of the inner Man." Let me explain.

The practical wisdom of learning "not to react compulsively" is taught extensively in Buddhism and Hinduism, and also – perhaps to a lesser degree – in the Judeo-Christian Scriptures. In Hindu Scriptures, such as *the Bhagavad Gita* and *the Upanishads,* as well as in Buddhist literature, you will see statements like: "While I am active, I remain inactive …."

A Westerner, unfamiliar with the Oriental teachings, might scoff: "What kind of gobbledegook is that; that's doubletalk?" Well, it may look like a contradiction – but it is not. The first half of that sentence refers to physical action, the second half to (a certain kind of) mental action.

So, what is it (the Oriental sage is telling us) that should first become, and then remain, inactive: his compulsive – residual – "savage" nature. And what is it that should become active: his rudimentary (partially developed – and therefore weak in most human beings) "Higher Self (for want of a better word)."

To put it all into practical terms: (as I have said earlier) Aggression worked well for the Neanderthal Man – but it *doesn't work well* for Man (in general) in the twenty-first century, and He *must* find a method to deflate it.

Not long ago, a man in his seventies was charged with road rage (I don't know what the legal terminology is for that phenomenon). In court he told the judge that the other guy had prevented him from being home on time to watch his favorite TV program (a feeble argument, wouldn't you say). She (the judge) – apparently inclined to use words sparingly – said: "After your ninety days, get a VCR!"

If the tragic old fool (as well as the back-window-smashing younger man) would have learned some basic common sense wisdom – not two days before the event but during several decades leading up to it – such as: "I can modify my old attitude of 'If you are a man, you stand your ground' to 'If you have the brains, you avoid altercations wherever and whenever possible,'" he would have made a modest beginning in – like St. George – "slaying one of the dragons" lurking in his subconscious mind. Is it practical, you ask? Well, did the man of a relatively advanced age (the "old fart" may be more appropriate) have a good time during his incarceration? He wasn't victimized by the other party, or vice versa, he – essentially – was victimized by the ferocious beasts of his own mind which under normal circumstances were locked in their cages but escaped to attack him in his moment of weakness. He never *intended* for any of that to happen, but happen it did – because he never learned to remain "inactive," or undisturbed, or "unflipable." If the mind isn't flipped or turned upside down, then the operator of that mind isn't flipped, and he or she has no inner reaction – and that has a far wider application than just in road rage incidents. In the beginning you may, like a good actor, have to fake it – showing no outer signs of the still present emotions while maintaining your composure – but over time it will, like anything else we practice, become second nature (inside and outside). Yes, that kind of learning *is* available to everyone, and isn't that so much easier and less hazardous than uselessly trying to convince the probably unreceptive other party, repairing a broken car window, getting arrested, getting injured, or worse?

Nobody can flip you off – even though you have no way of preventing another from sticking up his or her middle finger. The latter is outside of you, the former inside – and you have the absolute, unqualified power to keep the door of the mind closed to the offending party.

To remain undisturbed or serene at times like that – and ideally at all times – is literally "cool," because to get "all worked up" or "pissed off" is literally "hot," and hurts you far more than the other guy.

Oh, yes, we can – in spite of all the psychology teachings to the contrary – modify and improve our own character and thereby our personal destiny; and this is fact and not just preaching. However, if you – like the "tombstone mentality" in government (being reactive rather than pro-active) – postpone that process until a moment of crisis, your old second nature will seize control of your behavior at that moment (our devices are overthrown – Shakespeare) and the outcome will not be to your liking.

Is it practical, you ask?

Let the women be silent in the church

Since this chapter is about rejections of religion, I might as well tackle that one too. That well-known New Testament statement by the apostle Paul: "let your women be silent in the church," is a "neat" or "fun" one to "play with" for the non-indoctrinated freethinker. But how many feminists or even ordinary modern housewives appreciate the inclusion of that statement in the Sacred Book? Many church-going women take solace in the assurances of their without-any-other-choice spiritual leaders that this saying no longer applies to the female gender in modern times, which purely arbitrarily – and for no other reason than that they cannot make sense of it – pronounces a part of their beloved "Word" irrelevant.

So, does it have a practical meaning, or is it just further evidence that these ancient writings are nothing but "a bunch of horsefeathers"?

Well, one thing we do know is that the Middle-Eastern first-century writer was apt to use a lot of symbolism. Does that mean that I know that St. Paul only intended it to be interpreted as a metaphorical expression? That doubtful question may be relevant to a Bible scholar, but it isn't to me, because whether or not the writer was even aware of it, a well-fitting-basic-kitchen-variety-psychological in-terpretation is possible, leaving all other considerations nongermane.

Actions of the conscious mind, such as reasoning, decision making, evaluat-ing, etc., have often been identified as being masculine qualities; reactions – most-ly purely subjective – such as exuberance, anger, joyfulness, resentment, laughter and crying, are frequently – rightly or wrongly – categorized as being feminine mind characteristics.

For about a half a year now (I believe since September of 2002), the in-front-of-a-huge-TV-audience short counseling sessions by the catapulted-into-super-star-status psychologist, Dr Phil McGraw, has been challenging his guests to con-sider substituting a not-working-for-them behavioral pattern for a better one. In almost all instances, it is clear that the guest on the show (or the patient on any psychiatrist's couch) is emotionally not completely ready – notwithstanding their verbal assurances – to let go of their old ways of doing things. They have declared that they want to change – and that is why they are on Dr Phil's program in the first place – but there are forces that seem to be holding them back. They are the subconscious feelings proclaiming their objections to the uncomfortable decision to change. Could we symbolically re-phrase that to: "these are the wailing women (remember that we are now dealing with imagery and not people) who are mak-ing distracting noises in the sanctuary of the mind, which may prevent the men from deciding on (what emotions are incapable of doing) and implementing a new course of action"? Is that why these metaphorical women have to "shut up" so that the relatively fragile (masculine) decision has a "fighting chance." Oh how relevant that is to the human condition! At the literal level that statement is "pure baloney" – but as a metaphor it is astoundingly accurate, appropriate and meaning-ful. Aha, so it is advisable to dig a little deeper.

Horsefeathers or horse sense – which is it?

Familiar words in a wedding ceremony

"Therefore shall a man leave his father and his mother, and shall cleave unto his wife: and they shall be one flesh." These are familiar words from the Old Tes-tament (King James, Genesis 2:24) that are often used in a wedding ceremony. Did the author, whoever he was (Moses or somebody else), just inform us that at a certain point a young adult male will vacate his former living space in the parental home, go to the altar to say "I do," and live happily ever after with hopefully "the love of his life." If that is *all* it means, it is a bit of a waste of ink and paper – or more likely papyrus – because even my 6-year-old grandson knows about human arrangements like that.

On the same Dr. Phil show (I appreciate his helpfulness – but I hold no stock in any of his financial holdings), a young single mother solicited his help in trying to regain custody of her two children who were – after medical causes were ruled out – taken away from her by the State of Michigan. She had "lovingly stuffed" her young son with so much of the wrong kind of nutrition that he had "ballooned" to about twice the size and weight for his age. Over time his weight returned to normal in the care of a foster home.

After the example of "the silenced metaphorical women," it is not difficult to see the connection between the ancient wedding ceremony words and the plight of this modern woman. At a certain level she already knew – even before appearing on TV – that she had to make a decision (the young man in her mind preparing to leave the comfort of the parental home) to abandon the health destroying, distort-ed-maternal-instinct-driven behavior (the father [thought] and mother [emotion] of that dominant old mind pattern) towards her son *and* put a corresponding feeling-good-about-it emotion (the new wife) behind that new loving-my-son-with-better-nutrition decision (the husband cleaving to her), because without that the complete unity (one flesh) of these two elements of the mind, a relapse would be a virtual certainty. Oh, how closely that metaphorical interpretation fits! It is this absolute perfect match itself that confers authority upon that interpretation, without there being any need for a proclamation from a pulpit by an impressively attired, proper-ly ordained clergyman to confirm or deny its correctness. It is also not necessary to study ancient manuscripts in order to unlock some sort of secret code to an exalted revelation – and oh how some people love that deeply mysterious stuff – because the interpretation plainly and simply proves itself and that's all there is to it.

Unlike the physical event described in Genesis, that psychological process must be repeated many, many times during the lifetime of Man. The latter – and by far the more meaningful one – is also one of the principal methods with which He has an opportunity to advance in the evolution of consciousness process.

I repeat: "If the preacher would have explained these things in these simple – and yet profound, practical, and useful terms, would religion – considering the result-getting popularity of the Dr. Phil (living in a seven-million-dollar home) show – have become more popular than it is today, instead of uselessly remaining shrouded in mystery and cloaked in that unfortunate, decorated-with-dust-bunnies other-worldliness"?

No Religion – no Masters

I read those words on a bumper sticker at one time. The owner of the vehicle with that bumper sticker – like millions of others – clearly associated the idea of religion with the all-too-common notion of a Power or Celestial Authority over, or above Man. Of course, that belief or image comes from the Judeo-Christian belief-system. Whoever places such a sticker on his or her vehicle must be a member of the non-religious community who resents the ever-watchful looking-down-from-above Deity concept.

Now, because that higher-ranking "boss" image originates from certain Bible (mis?)interpretations, it is necessary for me to use other passages from that same book to show that such an image is only entertained by the less aware, or spiritually-less-mature type of mind. (Don't worry, being a person without absurd doctrines, I can assure you that I won't sneak one in through the backdoor. Keep reading, it'll make sense.)

There are many Scriptural passages that would qualify. One of the best ones may be found in Hosea 2:16 (King James), where it reads: "And it shall be at that day, saith the Lord, that thou shalt call me Ishi, and thou shalt no more call Me Baali." Baali means Lord, or owner, and Ishi means husband. So, here we have a female slave who becomes liberated and then enters into an equal partnership in marriage to her former master.

Now, because the Divine cannot change – I am the Lord who changeth not ... – the change or transformation took place within the individual mind, and only within that mind, and nowhere else in the Cosmos. This too is very much a part of the much-discussed evolution of consciousness process. This Bible passage is unambiguous (a great many of them are not) and therefore cannot be interpreted in any other way.

So, the man or woman who pictures a Master, or a King, or any kind of Authoritative Figure, in this dimension or any future dimension – "over" or "above" him or her, is still stuck in a "Baali consciousness." Most Christians and many others are stuck in it. However, when greater enlightenment comes, he or she experiences the physically formless (a mathematical necessity – otherwise it would be partial [forms by definition are limiting] instead of whole – and consistent with the most enlightened passages of the Judeo-Christian Scriptures) Supreme Cosmic Intelligence as a non-regulating abiding Presence at the core of their own being. It's actually very simple and non-mystical: the impersonal, non-initiating principle of mathematics walks with you wherever you go and is totally devoted – as if there were no other person in the world – to your using it, because it "lovingly (it never cheats you, disappoints you, or 'screws you over')" responds to you without fail every time you punch in a few numbers (your input) into your calculator (providing its battery is charged). Similarly, the impersonal Infinite Intelligence of the Cosmos will respond to you without fail with hunches and other communications of wisdom after your correct (and that's the key word here) input (providing that battery is charged too, namely, your charged expectancy) – because these two realities are natural and immutable "non-willy-nilly" principles. Is the principle of mathematics your master? The question is even quite silly, isn't it! To contemplate that a Cosmic Intelligence (or God) – or "His single *male* off-spring (yes, go ahead and scream here, feminist!)" – may be a master is equally silly! In spite of all the sermons ever preached, you were not made "to eventually piously sing His Praises in a heavenly choir," because there is *zero* evidence that there is a He to sing to. However, you'll probably feel – as most people do – a very natural (inner) urge to sing in the shower on a day when you really feel good about yourself, your life, and the Cosmos. That very real "heavenly" feeling can actually be generated whenever you choose, and at that moment – even if you don't remember the song's lyrics, can't carry a tune, and sound no better than a cross between a belligerent tomcat and a frightened chimpanzee – you've joined the only heavenly choir you'll ever "need." Wow, now it suddenly all sounds so simple. You know what: it was simple all the way along. That's why it's not a bad idea to stay away from babbling, prat-tling, jabbering, gabbling, chattering, and cackling fiddle-de-dee theologians!

"And when he was demanded of the Pharisees, when the Kingdom of God should come, he answered them and said, the kingdom of God cometh not with observation: Neither shall they say, Lo here! or, lo there! for, behold, the kingdom of God is *within* you (Luke 17:20-21)."

It cannot be stated in plainer terms. Not "with observation," because it doesn't have "a tinker's damn" to do with physicality – and it's a good idea to scrap that notion for all time. It's "within you" – or in your own consciousness! It's the real constitution of your own being – and this is not a doctrine, but a basic fact of your own existence. Not only – or even primarily – in a possible after-life, but while you're shampooing your hair and practicing your Elvis impersonation in the shower.

Physically outside as an incoming "Baali authority" image on the retina of our "current eye(s)" or of our "future eye" – or – "seen" and "experienced" by the inner eye, or the mind's eye – without images of pearly gates, flying angels, praying saints, judging saviors, and a personal fatherly deity to worship (all non-physical metaphors and thus the forbidden graven images of the Ten Commandments) – as an unfailing formless reciprocal (Ishi) Creative Intelligence and Power at the center of your own being. Take your pick, you can't have it both ways! The Buddhist understands that, but why does the Christian and the Jew have such a hard time with the Exodus-mandated concept of "imagelessness"? They don't seem to be able to get away from this picture of "holy halo-ed guys" … "up there"…. The fact that this word "imagelessness" doesn't even exist in the dictionary (you won't find it in Webster's unabridged dictionary, although the word "formlessness" is listed)) demonstrates its regrettable unfamiliarity. But that is what "a within kingdom" unambiguously calls for!

It is the "great Nothing" – look closely at that word "Nothing" … aaaahh, now I'm getting it: "No-thing"!! Thank Goodness, it's not Something … because Something is "stuff" … and "stuff" has *no primary creative power* … ONLY CONSCIOUSNESS or NO-THING DOES!

No masters over us – none – now or ever!

Too much art and not enough science in religion

Yes, that is exactly my sentiment (and, no doubt, the sentiment of millions of others on the planet): "Too much art and not enough science in religion (or, if you prefer, 'spirituality')." I'm not suggesting that "art" – or the personal element – should be "taken out of it entirely," but I do think it's high time that "science" – or the impersonal element – is "put in to it."

One must remember that all the major religions of the world – including Christianity and Judaism – were "born," or formulated in pre-science times, and even long before the Dark Ages. Moreover, the literature, which helped shape these (two particular) religious philosophies, was produced by men – their gender being very transparent in their words – some of whom would not have ranked very high in Abraham Maslow's self-actualization model or standard, or on the more general evolution of consciousness scale. If you doubt that, read the 25th Chapter of the book of Numbers in the Old Testament. The bishops and other church leaders in the early centuries – many of whom were reputedly illiterate – should have, while they were deciding on which books to include in the Bible, edited that inherited-from-the-Torah, written-by-a-zealous-bigot chapter out of "the Good Book – or *their* Truth Manual." It's probably too late to do that now, because few ecclesiastical authorities anywhere would dare – probably mostly for fear of incurring the "Divine Wrath" – to propose an amendment to the Holy Word. I – not being under any such spell – have a few suggestions for other amendments ... but how far will they go?

The spiritually timid one can rest assured that it *is* perfectly alright to bulldoze down a holy edifice already in ruins – in the mind only, to knock over *all* the worshipped religious images – in the mind only, to burn down a worthless shrine – in the mind only, and to slaughter all the sickly sacred cows – in the mind only. Nothing dreadful will happen, but you may hear a Cosmic sigh of relief (not to be taken too literally) – and thou mayest (for Pete's sake, don't look this up in the dictionary!) do likewise!

For these impossible-to-defend-in-modern-times, vile and repulsive expressions of what-is-supposed-to-pass-for-Truth alone, it is desirable to redirect our attention to science. Because we all have such privilege, I would like to offer an alternative hypothesis – which, of course, by its very nature is speculative – to the earlier-in-this-chapter-mentioned "religion is a property of the brain ..." hypothesis by Michael Persinger. Yes, I did say that I – never having had a personal encounter with these phenomena – don't understand the nature of the so-called spiritual experiences that he was trying to investigate. However, in spite of that, it is still possible to theorize "how these things may come about."

A good analogy is usually very helpful to the explanation. Most of us remember sitting in a movie theater being totally engrossed in what's enfolding on the silver screen. It's quite possible that you momentarily "lived vicariously through one of the characters in the story." Much to the chagrin of the mystically-inclined personality, who undoubtedly prefers that such designations remain reserved for so-called "spiritual experiences," it can be accurately stated that you were temporarily "in an altered state of consciousness" – because, while the body was sitting in the theater chair, the mind was in a make-belief world (I always like to strip the "other-worldliness" from these realities because they usually are far more common and ordinary than what many people think they are). Then the story ends, the movie is over, the lights in the theater are turned on again – and suddenly the released-from-a-virtual-reality mind "rejoins the body" and the redirected eyes notice all the to-be-swept-up-and-sold-the-next-day-again, non-virtual popcorn scattered on the floor.

Could it be that something similar was happening with (or in) Persinger's magnets-on-a-helmet subjects?

In earlier chapters, I have already alluded to the Supersymmetry theory, which is part of the larger String theory of the universe – now strongly endorsed by non-empiricists, who are not in the habit of surrendering their good judgment to unbridled fantasies. As I have explained earlier, there has to be another "balancing" reality of yourself in the next higher dimension – and perhaps in all of the other ten dimensions that the Supersymmetry theory calls for. That those realities of the human being – as well as everything else that exists three-dimensionally – interact between dimensions is a virtual certainty.

I like to stop here for a moment and do a little fine-tuning on the mind of the unrepenting empiric who feels that, because he can't see, hear, smell, touch, and taste these realities – they cannot exist. Are there TV signals, radio signals, cell-phone signals, cosmic rays, ultra-violet light frequencies, etc., etc, etc, all around us, and are they detectable by our limited five senses?!?! Duh, …! You should thank your lucky stars for not being able to see (with the naked eye) infrared radiation being emitted from your girlfriend's pretty face, because that would certainly cause a break-up. Oh that pitiable and doleful mind that is trapped ….

After a human birth, consciousness is focused through the five senses on things three-dimensionally. If that weren't so, we couldn't function as human beings in our world. But as I have proposed earlier, it may well be – in this instance, I'm prepared to go so far as to say: highly likely – that the elusive core awareness (see chapter 2) exists primarily in one of the other extra-dimensional, or non-three-dimensional realities (or universes). If that is so, then core awareness "looks" from its right-next-door – or probably even closer: right-in-the-middle-of-it – extra-dimension into the familiar three-dimensional world, similar to the movie watcher – largely withdrawn from events on the theater floor – looking at the screen in front of him or her. Could it be that the spiritual experiences – such as seeing "a larger Self" – induced by the magnets around the heads of Persinger's "human guinea pigs," are nothing other than momentary interruptions of the looking-into-the-three-dimensional-reality process so that core awareness' diverted focus "sees" its extra-dimensional reality temporarily, similar to the "attention return" of the movie watcher from the screen to the theater floor. That is not exactly a hocus-pocus, falderal-de-la hypothesis – or is it?

In the meditator it may be a chosen interruption or withdrawal – where with Persinger's subjects it would be a forced or involuntary interruption; and that may be the only difference.

Michael Persinger's argument that his "magnet induced spiritual experiences prove that religion is a property of the brain, and only of the brain …," is weak for the following two reasons:

1) His "brain only model" is dying because – as has been explained throughout this book – the Randomness model of the Universe is untenable.

2) If these "spiritual experiences" were "brain only phenomena," the experiences themselves would have been different.

Reason number one doesn't require any further elaboration here, but reason number two may be best explained with the example of a scientist looking through an electronic microscope. If the hi-tech focusing mechanism in the microscope fails and the picture becomes fuzzy or scrambled, he or she will pull the head away from the optics and refocus the eyes on something else in the room. The malfunction in the machine caused a cessation of an action (the studying of whatever was under the microscope) similar to the above-mentioned movie watcher's "return to the theater floor." But the magnet induced interruption of "the normal experience" of Persinger's subjects should also have produced "a scrambled, or a very confusing experience" for the subject instead of a pleasant and lucid larger Self experience, because the magnetic pulling action is not unlike what happens inside a VCR when a tracking misalignment occurs. When a video tape, which is also magnetized, begins to run off track, a distorted image with lines and "snow" is seen on the screen, which can only be corrected by a tracking adjustment. So, a "magnetic pulling" inside the brain should have presented *a distortion of the normal experience* to the awareness faculty of the Persinger subjects, from which there is – unlike the scientist withdrawing from the microscope – *no escape* in the "brain only model." But if consciousness can – voluntarily or involuntarily – look away from three-dimensionality and redirect its awareness to the extra-dimensional, or supersymmetric reality in which it primarily resides, only then can whatever crystalline images be produced on the screen of the human mind.

We might as well get used to the fact that the Universe is far more complex than what empiricism believed it to be.

The remote-control car-in-a-primitive-village syndrome

The extreme skeptic and the orthodox scientist may be suffering from what I would like to call: "the remote-control-car-in-a-primitive-village syndrome."

A twenty-first-century "civilized" young boy decides to have some fun by playing a prank on primitive, "stone age" youths in an isolated-from-the-modern-electronic-world village somewhere deep inside a jungle. He found a way to hide in the trees at the edge of the small village from where he can observe the movement of his remote control car. He races his gadget around the village square where the fascinated youngsters are trying to catch the what-their-brain-tells-them-must-be-a-never-before-seen animal. Of course, it's impossible for them to conceive that not all of the reality of what moves in front of them is "locked inside" the buzzing "animal" – and that the "primary reality" of the car sits in a tree a little distance away.

The erroneous conclusion after a split-second consultation with their brain memory data bank is completely understandable with these primitive young men. But what is the excuse of the orthodox scientist? "I can't handle another reality shift (or paradigm shift)"? He or she must have heard of dozens of them in his or her lifetime! That he or she – as a "non-hallelujah," cautious, "healthy" skeptic – doesn't embrace the new idea immediately is only evidence of his or her good judgment – but that he or she, without giving it a second look, "outright rejects" it, points to a syndrome.

Chapter Nine

The question of Personal Proof

Are we now not at the point where it is reasonable to assert that the likelihood of model 1 (chapter 1) being the correct model of Ultimate Reality is substantial? Everything that has been discussed in the first eight chapters of this book all seems to point in that direction.

At no time was the salesperson's method followed, which highlights and promotes the qualities and advantages of the product he or she wishes to sell (this sales approach is not necessarily, or inherently unethical, providing the customer still has access to all the pertinent information) while downplaying the importance of its drawbacks. It has been my attempt to explain and demonstrate rather than to promote and downplay. If something is true and based on identifiable and lasting principles, then it doesn't need to be defended or "helped," because such a reality is, on its own, quite capable of standing firm, and remaining so, during all the prodding and probing. No, it doesn't require a human defense, but it does need – what I have sought to do – to be pointed to.

But is it possible to get past this point of "substantial"? Is "rock solid" or "absolute" proof of the existence of an *a priori* Cosmic Intelligence available to Man? Universally or publicly: probably not! Individually or privately: most likely! I know I must explain myself more.

When we conservatively – even with a good dosage of healthy skepticism (not cynicism) – evaluate everything that has been placed on the side of the (figurative) scales of model 1 (see chapter 1) and consider the fact that nothing whatsoever was, or could have been placed on the side belonging to model 2, we can then make a legitimate claim to the conclusion that it is no longer reasonable to (still) think that model 2 might be the correct model of Ultimate Reality. That leaves us only with model 1 – but we still need "to rivet that down." However, should you decide that this conclusion is not warranted, then the rest of this chapter – and for that matter the whole book – will be of little use to you.

But I would be remiss if didn't share with you that, according to my own personal experience, it is possible to take this searching for evidence one step further. For some hard to explain reason, a direct personal strong indication (if we're not quite ready to call it "proof") of the existence of a Cosmic Intelligence and Its responsiveness to the searching individual, is far more satisfying than any other "outside" evidence we may hear or read about. What is presented to us by others, however solid or persuasive, is never quite the same as a personal experience.

The challenge that I would like to hold out to the reader is to see if he or she can personally duplicate the effects (of course, similar but not identical), which I have witnessed, of the application of what is now widely known as the Law of Attraction. There are quite a few motivational speakers (and there have been for decades) around the country and the world who point to that law in their lectures from time to time. The question is not whether or not it will work – in a Universe of Law and Order there simply cannot be any special dispensation for anyone – but rather will you, the reader, use and apply it in the way in which it works. This Law of Attraction is just as immutable as any other natural law; but precisely because it is a law or principle, it cannot be approached in a haphazard way if one wishes to see definite results.

How do we use this Law of Attraction? The answer is simple and direct: by using affirmations. Someone may say at this point: "Hey, wait a minute, isn't that essentially the same as the religious word "prayer"? The answer to that is: yes – and – no! It is a very definite "No," if a "petition or beseeching approach" is used; it is a definite "Yes," if the prayer or affirmation is accompanied by the conviction that, because we are dealing with a changeless and constant Universal Principle, the outcome is certain. Or, to put it mathematically: a specific unadulterated (never denied) affirmation + the conviction or belief that it will, or even must happen = the eventual (but inevitable) appearance of the manifestation. The "how it comes about" may be mysterious (beyond our capacity to trace it), or – if you like the word – "miraculous," but the "that it comes about" is not.

To many people, the words "miraculous" and "supernatural" suggest that a natural law was either broken or temporarily suspended. But there is nothing what-soever in the Cosmos which indicates that this has *ever* been, or even can be done. Unlike the law of gravity, or the laws of electricity, this Law of Attraction may not be (as) familiar to us; but familiar or not, it has to be a natural law or we cannot expect consistent results.

So, this is the point where science and essential religion converge. It is science because it is governed by law, and the best of religion has pointed to that law.

For the benefit of the reader with a religious background, who, because he has never had the working of this law presented in quite this way, doubts that "true" prayer will get definite results, I feel that I must quote one more time from the black leather bound book (I only want to do this sparingly, remembering that indiscriminate Bible thumpers gave me the creeps in my younger days; and also

179

because there used to be a middle-aged spinster in our neighborhood who forever said: "the Bible says …," so, we, irreverently, called here "Mrs. Bibels-ass." Well, that's what happens when you're into way too much preachin'!).

These quotations deserve their own paragraph because, in my opinion, they contain the most important words *ever* uttered on the planet. The following two unequivocal statements are from the highest Authority of the Biblical New Testament (King James): "All things, whatsoever ye shall ask in prayer, believing, ye SHALL (not may) receive" (Matthew 21:22) – and – "Or what man is there of you, whom if his son ask bread, will give him a stone? Or if he ask a fish, will give him a serpent? … (Matthew 7:10)."

These two statements clearly show that the kind of results we are reaping in life is – by law – directly, solely, and indisputably related to our own states of mind and has nothing whatsoever to do with the actions of an often-believed-in "capricious Deity." A great metaphysician once said: "Prayer doesn't change the changeless God – but it does change the transformable mind of Man." We would be far better off if we changed the word "supernatural" to "superbly natural," for in that way we would stay a lot closer to those two hard-to-misinterpret Scriptural statements. We experience what we *essentially* believe in and not what we are hoping or yearning for.

A reader may object with: "Whatever happened to your earlier promise not to promote a belief-system, or a doctrine? Isn't that exactly what you're doing now?" Absolutely not! I'm talking about a demonstrable and verifiable law which, if applied, can readily be proven by an experimentally inclined agnostic, an Eskimo, a Norwegian housewife or an Australian businessman. At this point we are *entirely outside the range of speculative belief-systems*, and – providing the reader can demonstrate these principles conclusively to himself or herself – have completely moved into the realm of pure science.

If the invitation to personally conduct this "inner" experiment sounds reasonable and promising, it can only be considered sheer foolishness not to go ahead and do so, in light of the fact that the workings of this Great Law include potential life transforming possibilities. No laboratory technician in his right mind would announce results before his experiments are completed. So why have an opinion in advance? You may be pleasantly surprised!

Remember Marcus Aurelius' admonition – which is so unnatural to the average man or woman: "Let your opinions lie still …."

I would like to share one of the remarkable personal results that I got – or more accurately: happened to me (because we don't "create" the results ourselves – "it is done unto us …") – with the application of the Law of Attraction.

While living in Sacramento, California, a little over twenty years ago, one of the bearings in the power steering pump of my Chrysler became quite noisy. Because I've always enjoyed tackling mechanical problems, I decided that it was time to put the pump on my workbench in order to replace that bearing. It turned out that I needed a special puller for that job. I had a delightful excuse to make space for it on my tool-board, but I wasn't quite ready – my wife was, unsurprisingly, even far less ready – to justify an almost $60.- expenditure from the family budget for this item. So, on a Friday night, while standing in front of my workbench and looking at a picture of that puller in a catalog, I affirmed that I would quickly acquire that tool without that uncomfortable extraction from our bank account. Early the next day, Saturday morning, the family got into the car to make the eight-mile trip over backroads to the Roseville auction. While my wife looked at the vegetables and the specialty cheeses, I followed the sign that read: "Antique and miscellaneous junque (I suppose that translates into highly collectible junk)." Under one of the folding tables, a vendor had placed a bucket full of tools. I noticed in a flash that the superficially rusted claws of a half-buried tool resembled a part of the puller

that I was looking for. So I pulled it out of the bucket, and at that same instant realized that I held in my hand the manifestation of the affirmation of the night before.
Being the natural negotiator (I am the odd-ball in my family) that I am, I asked the
gentleman: "How much for this rusty contraption"? His surprising answer: "Oh,
just four bits." So, in my euphoric state, I said: "I usually dicker a little more, but
I think it's fairer if I give you three bucks for it." Without saying anything, but
with that typical this-guy-must-be-a-fruitcake expression on his face, he opened
his hand, in which I dropped my three dollars, and thus sealing the transaction.
I cleaned the Cal-Van puller and removed the rust in my garage, and the look-
ing-like-new-again tool flawlessly pulled the bad bearing from the power steering
pump. For the benefit of the reader who doesn't know much about tools, it must
be understood that this specialty tool is rarely found in toolboxes of professional
mechanics, or even in shops for which they work. My youngest son, who has been
a Master auto-technician for almost ten years now, told me that car dealerships do
not want to their mechanics to be slowed down by the individual parts recondition-
ing process (the labor charge being as high as, or possibly exceeding the cost of the
new part). A new pump – rather than patching up the old one!

That specialty puller is as uncommon as the hammer and the screwdriver are
common – and therefore its appearance within 24 hours was remarkable.

I still have that tool hanging on my tool-board today. But I also keep it – without breaking any law of physics – in a far more important place than that. It's on
my list of remarkable manifestations or demonstrations. I've gotten a tremendous
amount of good use from it over the years, even though the metal object was only
taken down from the board no more than two or three times during all that time.

When such a manifestation comes to you, you most likely will cherish the
memory of it – and how it added to the conviction of the existence of an Infinite
Intelligence and Creative Power – much more than the acquisition of the physical

thing itself. Henry David Thoreau said: "That we have but little faith is not bad, but that we have little faithfulness is, because by faithfulness, faith (not a creed) is earned."

When you compile a list of these personal remarkable manifestations – I have written down 311 of them and added conservative statistical ratings to each over the past thirty years – no human being, even if he or she has a dozen Ph.D's behind the last name, and not even "the doubter" within yourself, can ever shake your awareness that only model 1 (chapter 1) can be the correct model of the Universe. I consider that list my most valued possession; and every time I look at it, I say to myself: "Yep, there it is, it is not only solid, it's even absolute."

However, I don't think it is helpful to tell the story behind each of my other 310 remarkable demonstrations, because that would take up unnecessary space and would probably be quite boring to the reader. Nonetheless, it seems to me that whoever possesses such a personal list, cannot help but feel, as I do, that he or she has arrived at that wonderful point when the Cosmic Intelligence and Its creative responsiveness has *totally* proven Itself to him or her. That man or woman can no longer be under the sway of the however impressively sophisticated, or ridiculously feeble-minded naysayers of society. The perfectly argued denial will blush with embarrassment when unassailable proof walks into the courtroom of the mind. The former doubts, hopes, and beliefs have been exchanged for science and knowledge; and all things to the contrary have become irrelevant.

One word of caution! It is highly advisable for the beginner to start using affirmations for small or relatively insignificant things or situations first. Why? Remember that we are working with a natural law and under that law you and I must furnish the correct input, or the unadulterated state of mind, before the desired "matching" physical manifestation can make its appearance. If we were dealing with a "sweet sugar daddy in the sky," that individual could change or override our imperfect request and supply us with "goodies" in spite of our disbelief, hesitation, denials, or low expectations. But a natural law doesn't do that. It creates according to the image – and the expectation behind it – that we entertain; and if that image is imperfect, the manifestation will be likewise. As you sow, so you reap. It is similar to the well-known computer slogan: "Garbage in, garbage out."

As in any other scientific endeavor, we must remember that, here too, "Nature only obeys us, after we *first* obey it" – which simply means that we first must study the workings of a natural law before we can use it to our advantage. So, the reason that an affirmation applied to a relatively insignificant thing, situation, or condition gets better and quicker results is that there are no anxious – and therefore opposing – thoughts and emotions that neutralize the power of the spoken word. Little or nothing is at stake; and because it's not a "big deal," there is no blockage or ob-struction in consciousness. I spoke the "puller affirmation" out loud, then dropped the thought, and didn't even remember the affirmation of the night before at the auction. But apparently there is something in us, or with us, that doesn't forget and is busy working with this creative process while we are doing other things. No serpent for a fish and no stone for a loaf of bread.

When we take our first piano lessons we practice on simple tunes like "Happy Birthday," and no piano teacher would introduce you – if ever – to a "hard to play" piano concerto by Franz Liszt until you have practiced for countless hours over many, many years. Similarly, we must recognize a principle of growth in the use of this great Law of the Universe. Am I suggesting that you shouldn't use affirma-tions on the "larger" challenges soon after learning how this Law works? Not at all! What I am saying is that after you've "bagged" a few "puller level manifesta-tions," you will have a lot more confidence in your own ability to use any affirma-tion with the required high level of conviction.

Because the subject matter of this book deals with the existence question of an *a priori* Cosmic or Universal Intelligence, it is necessary to explain why, so far, I have – in referring to this Intelligence – chosen to use the neuter pronoun of the third person, "It," rather than the male pronoun of the third person, "He." The English language, as well as most other languages around the world, doesn't provide us with a good, or right word to use in this instance. All of the three pronouns of the third person – He, She, It – are limited, inadequate, and therefore wrong. The Supreme Being behind all of Life is as much "human" as It is "dolphin" – and at the same time cannot be (limited to) either one of these "expressions" of Itself, because that would deny Its Universality. It cannot be locked into any form of Life. It is all of the forms – and, paradoxically, none of the forms at the same time. It is above all else Infinite Consciousness. But Man is also *primarily* consciousness (and only secondarily a physiological being), and – so is the (possibly equally intelligent) dolphin!! It (the Cosmic Intelligence) has "beloved human sons and daughters," but It also has, among others, "beloved dolphin sons and daughters." However, I do not blame the Man-having-dominion-over-all-the-Earth author of the first Bible book, who had most likely had never even heard of a dolphin, for not considering that reality in his writing. The Infinite expresses Itself through many forms of Life and not just the human one. I'm sure many Sunday-school-taught people have a hard time with that last statement.

"He," conjures up the "Man upstairs" or the "Old Man in the sky" image; "She," may bring to mind a picture of a "Celestial Fairy" – which is equally absurd (although I can fully understand why some feminists prefer to use "She"); and "It" may suggest that we are dealing with nothing more than another law of Nature, instead of with the All-Originating Life, Power, Wisdom, Love, and Life-giving-ness of the Universe. Yes, the creative manifesting power behind the spoken word must out of sheer philosophical necessity work as a natural law or principle – but we must remember that Life Itself is always infinitely more than any aspect of It.

How does one solve or overcome that third-person dilemma? By explaining what we mean a little more, as I just did. But I would really prefer to see another word created for this purpose. However, I am well aware of the fact that a – perhaps somewhat amusing – suggestion, such as "Hesheit" will never make it to Webster. Oh well, …!

So, the reader must personally decide which third person pronoun to use and what words carry the greatest "impacting" meaning, as long as it is borne in mind that it is never the words themselves, but rather the required consciousness of expectancy that we seek. And as the above quotation of Henry David Thoreau suggests: this consciousness of expectancy or faith can expand, and its manifesting power multiplied – consistent with the great general evolution of consciousness principle – through practice.

A good way to become comfortable with a word

In the early nineteen-eighties, a woman approached me after one of my lectures with the comment: "In general, I like what you're telling your audience, but I have a hard time with the evolution concept, and that we are supposed to be descendants of the apes." I had used the word "evolution," but I never said anything about apes that day. Nevertheless, I understood her "hang-up" and knew "where she was coming from." So, after a brief moment of searching for an answer, I said: "As you may know, the early Egyptians lived about five thousand years, or fifty centuries ago. On an average there are approximately five generations of ancestors in most family trees in each century. Five times fifty gives us two hundred and fifty generations of people who were just as human as your Mom and Dad, your Grandma and Grandpa, etc. Print a list of two hundred and fifty fictitious names and put "human being" behind the name, and you'll be cured from your feeling of discomfort. The apes weren't our ancestors, they were our very, very, very, very, very distant ancestors. However, if that still "doesn't quite cut it," go back to our

Cro-Magnon ancestor who, being a lot like ourselves, already walked the Earth at least a hundred thousand years ago, and put that "info" into your calculator and make another this-time-probably-ten-feet-long list. And, by the way, I'd be more embarrassed or worried – if that is what you've decided to continue to do (a little friendly sarcasm to make the point) – to have a character like Adolf Hitler in my family tree than one of those lovable chimpanzees."

She later confirmed that her uneasiness had disappeared as a result of this simple contemplation.

Our hang-ups are in the mind; the solutions to them are there also.

Clearing up the most supreme of all the widely-held first-class misconceptions

"Man, you've got a nice collection of superlatives in your heading, why didn't you add a few more," somebody may react sarcastically. Well, I didn't use them "just to be cute," they are there to draw attention to the fact that if you get nothing else out of the book, the understanding of the following alone is worth your moderate investment.

There are those members of society who are shouting from their roof-tops that the most important thing (and sometimes: "the only thing") we should do for ourselves to cure what's ailing us is: to swallow our daily amount of vitamins, to eat certain foods while avoiding others, to follow a certain exercise routine, and to have our regular medical check-ups, of course, followed by a bank-account-draining visit to the local pharmacy – and in the absence of unfortunate genetic predispositions: "All shall be well." And then there are those who say that "it's all in the mind," and all we need to do is "change what's there," and all the "other stuff, such as proper nutrition, is more or less irrelevant." Those disparate positions are one of the many polarities in society. Who to believe?

The first group of individuals – a lot of them are from the old school of medical practitioners – disbelieves, or just ignores the fact, that a state of mind is essential in determining what will happen next in the physiological system of a patient, under medical care in particular, or healthy human beings in general. That can easily and readily be disproved. Pump an ailing man or woman – who feels that he or she no longer has any reason to live – full of vitamins, medication and what have you, and life still withers away; bring Man's best friend – with a wagging tail and a tongue ready to lick you all over – into the patient's room, and there is a pronounced resurgence of life, because the dog – especially an adorable puppy – "injects" the patient with the potent vitamins of joy, laughter, companionship, and unconditional love. Doctors all over the world are beginning to recognize that. More than a few are now allowing, or even actively encouraging the presence of good-natured canines – even without first sterilizing them from top to bottom – by the side of the bed of the trying-to-recuperate human being. They are natural healing facilitators and accelerators.

There is a neat little poem that reminds us why that is so:

> *Man may smile and bid you hail*
> *Yet wish you at the devil*
> *But when a good dog wags his tail*
> *You know he's on the level*

Using a good analogy often leads to the "Aha, now I see ..." experience. We all hope that when we take our car to a repair shop for servicing, the work will be performed by an experienced and trustworthy mechanic (now often called auto-technician). Most of us would be *far more* interested in the man's (occasionally a woman) level of competency than whether or not most of his tools in his collection were made by MAC or Snap-On (the ultimate and most expensive tools to the mechanic). Would *any* savvy consumer choose the disinterested left-handed klutz with thirty thousand dollars worth of such tools in his box over the eager "Midas-touch" master technician, who – because he has a young family to support – has only been able to build up a modest tool collection with the words "made in China" engraved on some of his screwdrivers, pliers, and wrenches.

Aaaaahh, so here too, it is the "mind" and not the – however fine in quality – "stuff" we're *primarily* looking for.

So, the man or woman who only consumes the finest or healthiest foods, but who neglects, or is unaware of the need – or stronger yet: the necessity – to put his or her "emotional household" in order, may be compared to the incompetent mechanic with the high-grade tools. In most instances, his or her health will still be poor or marginal, because the system is sabotaged by a subconscious mis-programming and the release of – manufactured by the body itself – harmful chemicals. These are now well-proven scientific facts and are no longer debated within circles of "the aware." The biochemical components in question – if extracted from the blood of an individual in a sustained state of anger or rage – would kill a laboratory mouse in a few minutes, or less.

And then we have the type of individual represented by the late George Burns, who is probably remembered by every adult in America as the lovable never-without-his-stogie comedian and movie actor. He was asked in his eighties if he was concerned about the preservatives in his food, to which he quipped: "Listen, at my age I need all the preservatives I can get." Was his lifestyle healthy or unhealthy? Well, his dietary and his ten-cigar-a-day habits may place him in the latter category, but some of his well-known mental and emotional habits force us to conclude that, on balance, he did live a healthy living-to-become-a-centenarian life. George loved his work – if you can call it that (it's more like play) – and obviously loved humor. He also had definite goals even at a very advanced age. All three of them good-chemicals-producing components! So, may we say that old George was like the analogous competent master mechanic with a not too terribly impressive set of tools? We all know from experience that – unlike the incompetent one – such a technician, in general, still gets the job done. However, the well-equipped, equally competent mechanic has a distinct advantage over the able man without the proper quality tools. So there we have it! Yes, the quality of the food we consume is important – but the governing (subconscious) mind in us – with its various states – which appropriates the substance, is, without question, *preeminently* important.

Did that take care of some big-time misconceptions?

A quick return to the weigh scales

Isn't it odd, that wholesome emotions "just happen" to produce beneficial and – even healing – chemicals in the human body? If it would have been up to Randomness, could "doing the right thing" – which, without question, makes every man, woman, or child feel good – have accidentally (wrong biochemicals) resulted in making the person feel either nothing or – pay close attention to what's coming up – "downright rotten"?

A little consistency here, model 2 supporter!

Conditioning: the glue of a creed

There is a special reason why I waited until this chapter before addressing the psychological principle of conditioning. I first wanted to show that scientifically verifiable, tangible results – duplicable by the reader – may (also) be obtained from the application of the Law of attraction with a non-indoctrinated, non-conditioned approach by a freethinker.

Any elementary book on psychology will deal with the principle of conditioning with examples, such as Pavlov's dogs, etc. I don't think I need to spend much time with that. It suffices to say that conditioning is a (mostly subconscious) learning process – but it is learning through repetition or training (like brainwashing) and *not* conscious learning through debating, reasoning, and critical evaluating – and definitely *not* discovery through contemplation. Most religious beliefs in most people are a product of conditioning (although the originators obviously formulated them after a period of reflection or contemplation). That means that conditioning is the glue that holds the shape of their belief together – even though the elements or components of the belief may not be well-fitting building blocks that would allow the structure to remain standing without the bonding power of that glue. Now, it is much easier to see – precisely because of the non-self-examining nature of conditioning – the flaws in the building blocks of "the heathen (somebody with a philosophy other than your own)" than the irrationality in the ones in "the heathen to your heathen (meaning yourself)."

There is an amusing story that illustrates that: "A western man put flowers on the grave of a loved one. On a grave a few feet away an oriental man placed a bowl of rice for his dearly departed. With a smile that betrayed ridicule, the westerner asked: "When do you think your friend will rise up to eat his rice?" Without a trace of indignation, the oriental man replied: "Same time your friend will get up and smell flowers."

The unexamined life isn't worth living (Socrates) – and neither is the unexamined thought worth thinking. Blessed is that man who recognizes the – sometimes pathetic – folly of his own conditioning ....

No doubt, millions of people in this part of the world consider it offensive to dissect the tenets of orthodox Christianity and the core of its belief. Well, is it taboo to disassemble it, examine the parts, and if found to be in good working order, reassemble the unit? But suppose we discover that the parts or building blocks are ill-fitting and cannot properly be reassembled without the "tacky" adhesive of thoughtless conditioning – then what? Ask yourself the basic question: "If a fifty-year-old man, for example – who had never heard of the Christian Faith and is therefore totally unconditioned in reference to its theology, and also has no alternative religious philosophy of his own – were asked, if he thinks it's reasonable for the Creator of the Universe to demand the execution of his only son as a ransom payment for the "screw-ups" of His wayward humanity – would he answer: "Of course it is"?!?! The answer is obvious – but he probably would have given that answer had he been religiously conditioned and remained so during his half century of living. That violating-all-common-sense-and-decency belief cannot stand up without that sticky stuff holding it together!

The early Christians, known as Gnostics, who obviously were much closer in terms of time distance to the first-century Gospels, were far more interested in the teachings of Jesus of Nazareth than in his crucifixion. The shift of emphasis didn't take place until the Dark Ages when the story of a suffering and unjustly condemned man appealed to the masses who had to deal with daily hardships and misery themselves. Now, I'm not suggesting that the crucifixion and the resurrection stories do not have a meaning other than the just-mentioned absurd one, but I don't want to get in to that for two reasons: 1) I made a promise not to get involved in theological speculations, and 2) I am, like the Gnostics, *far more* interested what can be scientifically done with what the carpenter's son had to say – such as the above-quoted "belief manifestation principle."

I want to till the soil of the patch of land I have in front of me – and don't want to waste my time dreaming about real estate up yonder.

P.S. Because it is not my present moment intention to write a lengthy exposition about the subject of the Law of Attraction or the Law of Mind, I would like to refer the interested reader to the excellent – scholarly written – book by Thomas Troward, *the Edinburgh Lectures on Mental Science,* published by Dodd, Mead

& Company, New York (printed before the issuance of ISBN numbers). Should the reading of the book prove to be too difficult for the beginner, any of the books by Joseph Murphy, Emmet Fox, Ernest Holmes, Ervin Seale and many others will be helpful in expanding the understanding of Man's relationship to the Greatest of All Laws of Life.

Chapter Ten

Conclusion

Because there is such an abundance of extremely-difficult-to-explain-away evidence favoring model 1 (chapter 1) – no doubt only a tiny portion of it is set forth in this book – the question of why did people of the caliber of a Carl Sagan, consciously or subconsciously, "choose" model 2, puzzled or even haunted me for quite some time. From a choice of merely two possibilities, one does not – in the absence of any irrational motivation – "embrace" the more unlikely one of the two. A well-designed impartial computer program, fed with all the relevant data, would certainly – because of incalculably high odds against model 2 – have selected model 1 as the correct model of Ultimate Reality. It made me go back, particularly to the identical twin phenomenon (discussed in chapter 2), over and over again to check the correctness of my conclusions.

Although the possibility that I might have overlooked something is not non-existent – I am not so conceited to claim infallibility – the case for my inference of one-individual-being-inside-two-different-bodies-simultaneously in the "brain only" model, which is of course impossible, appears to be rock solid. Furthermore,

I completely fail to see how one can look at the reptilian egg odds (chapter 6) and not be a little shaken in, or a little less certain of one's atheistic convictions. While reflecting on the above-mentioned question, I decided to list the possible reasons for that philosophical position:

1) Insufficient intelligence
2) Habits of thought, including the habit of not questioning or examining one's own thinking; in other words, the slothful non-Socratic attitude
3) Having become more than a little "cocky" about the extent of human knowledge
4) Never having been properly introduced to the relevant evidential facts
5) Intellectual dishonesty; or refusing to look at the evidence with an open mind
6) Being turned of by, or even disgusted with religion, religionists, God, etc.
7) The it's-fashionable-in-a-certain-environment element
8) The public persona phenomenon
9) Standing too close to the tree

Going down the list, it is quite clear that insufficient intelligence cannot be the reason that Carl Sagan became an atheist (or at least he expressed himself as such) and Bertrand Russell a self-pronounced agnostic. Both of these men were brilliant. It is also probably true – I have never seen any statistics on this – that most atheists have a higher than average IQ. The Mensa organization has (or had) a disproportionate number of them among its members.

Now, it is certainly nothing other than a pure – but odd – habit of thought to think that "zillions (the choice word, when all we know is that it must be astronomically high)" of mutations – considering their now known astonishing complexity – all happened to "roll in" at the appropriate time during the evolutionary process while the lovely "Lady Luck" was at the helm of the ship. This closed-loop thought pattern apparently prevents one from remembering that her – yes, women drive their cars as well as men, but not this one – erratic steering makes a total shipwreck a certainty. But the ship is still afloat and life is still here. Is it really reasonable to think that under her watch even one mutation could have "popped up"?

Someone said at one time – in reference to the Pure Chance notion – that it is inconceivable that a bunch of Lego pieces, placed in a closed box, would even in a billion years become a recognizable toy, such as, for example, a fire engine, as a result of continued shaking (assuming, of course, that the force is sufficient to attach the pieces to each other). But the complexity of a fire engine pales in comparison to the complexity of the average mutation. Yes, in the process of the shaking some of the pieces may well *accidentally* attach themselves in a correct position. However, aren't the odds astronomically high that those pieces would separate again long before the arrival of the fire engine?

So it *is* a habit of thought, because from an unbiased perspective, there is plenty of room for doubt, which – if the other factors on my list of nine didn't exist – should, as a minimum, give an atheist enough pause to consider "converting" to, or reducing it all to agnosticism. Agnosticism is intellectually reasonable but – sorry – (pure) atheism is not.

The well-known agnostic, who has made a guest appearance in this book a few times, Lord Bertrand Russell, explained in his book *Why I am Not a Christian* why he was more than "a little ticked off" with the Creator in the following lines: "The world in which we live can be understood as a result of muddle and accident; but if it is the outcome of deliberate purpose, the purpose must have been that of a fiend. For my part, I find accident a less painful and more plausible hypothesis (page 73)."

"The world in which we live can be understood …"? You mean totally, Lord

Russell? A little exuberant presumptiveness? I don't think that I suffer from a severe case of diffidence and a low capacity to comprehend, but I wouldn't say that I understand more than a tiny fraction of everything that is "out there." The old adage: "The more you know, the more you become aware of all that you don't know," still rings true today.

His plausible hypothesis is my model 2 (chapter 1) of course. The eminent philosopher looked at life and obviously didn't care for much of what he saw. But model 2 has been looking a little pale lately and is running a fever. The first nine chapters of this book say that model 1 is the more plausible one, but here the esteemed fellow-pipe-smoking British thinker, for whom I have a genuine respect, is leaning towards model 2. A stalemate?

In order to get past that stalemate, I would like to introduce another hypothesis. I cannot prove the reality of it anymore than Lord Russell could prove his muddle and accident worldview. However, if – as other thinkers have suggested – this lifecycle is very much like a certain grade level in the larger school of life, and there are higher or more enlightened dimensions, not unimaginable in an intelligent universe – and no philosopher can exclude that possibility – then the picture changes dramatically. When we walk into an old school building where they have kept some of the papers of former students, we could justifiably say something like: "My goodness, look at some of the dumb mistakes they've made in their grammar and calculus," thereby momentarily forgetting that some of these former students have progressed to the level of doctors (barring their abominable handwriting, of course) and research scientists.

Not long ago, there was a documentary on TV about the life of the Neanderthal Man. At one point, a few men snatched away a young female from another group – for the purpose of adding greater genetic diversity to their own group – in a rough and "feminist unapproved" manner. But before we begin to condemn the stocky, broad-nosed humanoid for his actions, we must realize that being out in the open was dangerous, and therefore everything had to be done swiftly, that the getting-to-know-you courtship rituals had no evolutionary significance at his stage, that there were probably no language skills with which to negotiate anyway, and that restaurants with fine silverware for a romantic evening wouldn't be there until tens of thousands of years later.

So, the very same actions, which we rightly call "appalling" today, were necessary and therefore not "wrong" in former times.

(I am aware of the fact that I have already alluded to some of these ideas in earlier chapters, but I consider them of such great importance that I have decided to take the liberty to present them again in a slightly different form.)

Now, if the findings of Dr. Ian Stevenson (see chapter 5) prove to be correct, then we shouldn't be surprised that not all of the mind patterns of the savage, which were once extraordinarily helpful for his survival, have been – because of all the "baggage" that was carried along – extinguished in modern man. It would then also be perfectly asinine to blame a Cosmic Intelligence for the existence of the grade level – in which we find ourselves and out of which none of us have yet graduated – where the semi-Neanderthal also has an opportunity to learn and grow. Yes, he *can* "cast off" the old man, but it must be done *by him* in an incarnation – and apparently not even the Infinite Intelligence Itself can bypass that fact. To consider the reality of all these things in *that* way is far "less painful (using Lord Russell's own words)" to me than the "nothingness" and the "bleakness" that the philosopher's plausible hypothesis has to offer. If I may step out of my figurative three piece suit and put on my figurative pair of jeans for a moment, so that I can express it in the for-respectability-purposes-modified-into-a-softening-equivalent of a now common vulgar expression: " ... for that one really siphons!"

So, what about the young Neanderthal woman and all the other "victims" – the great stumbling block for Bertrand Russell as well as for Mark Twain (see chapter 8) – of the world in general? Doesn't that disgrace tarnish something – "somewhere out there"? For isn't it true that the helpless victim has no choice but to take the abuse from the more in control and more powerful victimizer? Perhaps not! If there is something missing within the victim's consciousness that could have, if it were present, prevented the abuse – if not entirely than at least the prolonged abuse – and maintained harmony, then we no longer have a reason to shake our fist at the Universe, as the two just-mentioned gentlemen did in their own way.

In the seventeenth, eighteenth and early nineteenth century, the "white savage (as a Caucasian I can label him so)" sailed to Africa to round up the, as he saw them, sub-human blacks that he could catch, to take them back to America as slaves. Many children in elementary school have already learned something about that extremely dark part of our history. However, what is not as widely known is the fact that serious attempts had been made to try to enslave the Native American, or North American Indian. After all, they were already here, and they were – according the perverted white logic – just as "sub-human," and thus the high cost of importation might have been avoided. Surely, the average black man had a stronger physique than the average Native American, but the latter could have been used for lighter menial tasks; besides the self-righteous white barbarian had no problem with the idea of enslaving members of both races.

But – the proud Indian wouldn't submit. They knew so well within themselves that the Great Spirit had not placed them on this Earth for the purpose of being owned by other human beings.

No doubt, some of the braves entered the Eternal Hunting Grounds prematurely as a result of the repeated torturous and diabolical attempts to break their spirit and beat them into submission, but they had already decided – perhaps with the aid of their strong religious beliefs – to choose death over slavery. They lived with their ultimate resolve by the words of the orator and statesman, Patrick Henry: "Give me liberty or give me death."

Then the abuse stopped, because dead men do not make very good slaves. And even though, as history shows us, the Native Americans were horribly mistreated in other ways, none of them – with the exception of some indigenous Indians under Spanish rule – were ever owned by a white man.

I am aware of the fact that there is some controversy over the reason why the North American Indian was never enslaved. The story that I heard in the nineteen-fifties was contradicted a few decades later. Some say that it had little or nothing to do with their resistance, and that it was solely based on their inability to endure the forced labor. But that negates the indisputable fact that the Indian in captivity was far more prone to escape, far less willing to resign and accommodate, and far less trustworthy to his captors than the black slave. There was a totally different mind-set; and that must have been a contributing factor.

Well, since none of us were there three hundred years ago, we all have an equally difficult time separating fact from fiction. Besides, if you doubt the historical accuracy of the Indian's refusal to submit, turn to the courageous, non-compromising philosopher Socrates, who – although he easily could have avoided it – drank the hemlock rather than yield to the Truth-silencing demands of the Athenian aristocracy.

So, the truth that the Great Spirit or the Cosmic Intelligence does not "give" ownership of one human being to another, is *in itself* not enough. It must be known and become deeply embedded into the total mentality of the individual before it can become fact and circumstance. That conviction was present in the North American Indian and more or less absent in the African tribesman – for had it been there – slavery couldn't have existed, or, as a minimum, would have been a much greater "headache" to the slave owner. But there is zero blame attached to that absence. None!

The good news is, that there is little doubt that many modern black men and women "own" much more of that desirable inner quality than their ancestors from Africa. For example, black people, such as Oprah Winfrey, Bill Gosby, and Colin Powell – who probably consider their blackness just as incidental as I do my white-ness – do not feel within themselves that they are second-class citizens, let alone victims, and – according to the law: "as within, so without," which no white person can naysay – for that reason experience in their outer sphere the exact reflection of what they rightly know themselves to be within. They therefore remain relatively immune to the attitude of the occasional "redneck from the smelly swamps" who chooses not to treat them as his equals because of their skin color. But then he (generically) – being without the understanding of the principle of unconditional love and acceptance – will only treat "his own kind" with respect as long as they conform to the "mold of his expectation"; and, oh, the terror of his wrath once they dare to step outside that mold.

The spirit of the earlier Native American was like the spirit of the wolverine and the spirit of the bull moose. The solitary wolverine is the most ferocious and the most powerful for its size animal on the North American continent. It is totally untamable and it can therefore never be domesticated. It will not submit to the wishes of Man. We will never see a wolverine on a leash; and we will never see a rodeo champion ride "his" bull moose into town; even though his antlers have spikes that might work fine as handle bars.

So, the victim can unlearn – as difficult as that may be – to be a victim, just as the victimizer can do his equally challenging unlearning; and modern psychology can render invaluable assistance to those who want to get away from those lowly states. It's do-able!

This story of the Native American, which illustrates how he avoided being enslaved, is a small but definite example of why we are wiser to look "inside" or "within" our own consciousness for the explanation of why we have the experiences – the circumstances, conditions, etc. – that "come to us."

One doesn't have to look far and wide to see the evidence of this "victim consciousness principle" in modern life. On the highly popular "Dr. Phil (already introduced in chapter eight) show" on the CBS network, the psychologist – with his wake-up call approach – reminds the victimized guest, that if she – or occasionally "he" – would have had a higher level of self-esteem before the abuse, coupled with the realization that other choices (in some cases the submissive role that the victim unwittingly played was even staunchly defended) actually *were* available, her experience would have been completely different. So, the victim problem is correctable – even though that may "take some determined doing." That fact weakens, or even cripples Lord Russell's "victimizer and victim argument."

Once such a principle has been appropriated, or mastered, then – and not until then – can progressively higher principles come into view.

So, because we do have the unqualified power to modify, rearrange, and get rid of what shouldn't be there, while – like the brave – fiercely holding on to what truly belongs to us, we have every reason to be optimistic about our tomorrows if we are as diligent in exercising that power as many of us are in keeping track of the score at the Superbowl. The latter is just fun and momentary, but the former brings us ever so much closer to the more ideal man or woman that we are capable of becoming. Yes, I know, the much-maligned preacher (I know, I haven't been all that nice to him) may have said something similar – but never mind him – it's reality and it works!

Even though I consider myself a diehard optimist, I have no illusions about the fact that the very best and most solid evidence is not always going to "win over" the entrenched mind to a new insight. When the needle got stuck in a groove of a record back in the old days, it would wear down the groove so deep – if the tone-arm wasn't lifted from the record in time – that the record became unusable because the needle could no longer get out of that track or groove. When one "hangs out" in an environment of like-minded, biased individuals, and, in addition to that is widely or publicly known for one's worldview, a change of heart becomes nearly impossible and is virtually unheard of. The Pope cannot become a Buddhist in this lifetime (without getting millions of people more than a little excited) – I am not suggesting that he wants to, or should – and Billy Graham cannot enter the Mormon Temple or the holiest places in Mecca for the same reason. The public persona has lost – whether that is good or bad – a certain amount of mobility and freedom in the evolution of consciousness process. John Smith, the postal worker of Timbuktu, can convert from Catholicism to Methodism without many people caring, but the "top" Catholic cannot. Could that public persona idea have applied to Carl Sagan and Bertrand Russell as well?

When one becomes totally absorbed in a specialty field of research, in other words, is completely preoccupied with any other narrow aspect of life, there is a real danger that one may lose the ability to "think outside the box." In my youth, I heard the expression a time or two: "Stand too close to a tree and you can no longer see the forest." A modern parable that I heard at one time – I have no idea who originated it, so I cannot give credit to any source – expresses that same idea so well. A few blind men were asked to examine an elephant. One man put his hand on the tail and said: "I am now holding a rope." The next one touched the side of the animal and said: "This is a wall." The third one felt the leg of the elephant and said: "This must be a tree." The fourth one put his hands around the trunk of the pachyderm and said: "It feels like a fire hose."

We have certain experiences in life, but by the time we are ready to interpret an experience, the (subconscious) mind has already, in a flash, added "stuff" from our vast memory data bank, coloring, or even distorting the interpretation accordingly. That is the reason why perfectly sincere people can give completely different eyewitness accounts in a trial. They were a witness to an event, but – without realizing the workings of this internal distortion process – rendered a personalized and altered version of what the faithfully recording video camera or camcorder would have reported. Some people in our court systems are now beginning to recognize that fact.

It takes a fair amount of internal discipline – not often welcomed by the average individual – to strip away that colorization so that we may see more clearly. Were Carl and Bertrand standing too close to their tree, and as result remained unaware of their own – present in all human beings – internal distortion tendency?

A peculiar habit of thought was exhibited by a brain researcher who apparently had been "hugging his favorite tree" and couldn't let go and step back from it anymore. He seriously thought – he might have said it in jest, but it didn't sound like it – that if there were a Cosmic Intelligence, there would have to be a giant brain "out there somewhere." That's a fine example of an "inability to think outside the box" thought pattern. It assumes, or even insists that an *a priori* Intelligence *requires* a brain. However, if from the day we were born, we – like our early pre-science ancestors – had never heard of the brain being the organ of animal and human intelligence, we could then quickly and easily accept the concept of "independent from the brain," or *a priori* (that which comes first) Intelligence. Now, like the brain researcher, I am fully cognizant of the fact that I don't know anymore than he does what *a priori* Intelligence is (it may never be [fully] knowable) – but that is wholly different from denying that it does or could exist. But why would that particular "not knowing" be uncomfortable? We don't yet know what the elusive quark is in the atom other than the fact that it is a mathematical necessity. And why should our ignorance about that fact be any less disconcerting? Another habit of thought maybe?

I have often asked myself the question why intelligent men, like Carl Sagan and Bertrand Russell, have – in their reasoning – found it necessary to move from one extreme end of the intelligence scale to the other extreme end. If one feels that the notion of *a priori* Infinite Intelligence *must* on philosophical grounds be rejected, why embrace the Zero Intelligence model, instead of opting for something in between these two extremes. With a certain sense of amusement, I will say that it is far more reasonable to think that Santa Claus and his elfs created – and thereby being responsible for the occasional unintentionally manufactured Frankenstein – the Universe, than that it was put together by the one that cannot even be placed in the "Dumb – Dumber – Dumbest" range. In order to even make it to the "Dumbest" category, there has to be *some*, however little, intelligence. Randomness' application to be admitted into that category would have met with outright rejection for simply not being smart enough. One must have – what I like to call – "an astounding and almost superhuman improbability tolerance" not to see the rationale of that.

Bertrand Russell uses expressions, such as: " … at the expense of Omnipotence …," thereby clearly indicating that he had a problem with the concept of an all-powerful Creator, because such a Creator "should have done better" and couldn't have created the "Hitler and Stalin" type of human devils. His exact words on page nine of the preface of *Why I am Not a Christian* are: "Apart from logical cogency, there is to me something a little odd about the ethical valuations of those who think that an omnipotent, omniscient, and benevolent Deity, after preparing the ground by many millions of years of lifeless nebulae, would consider Himself adequately rewarded by the final emergence of Hitler and Stalin and the H-bomb."

But if it represents a dilemma, why then not "turn it all down a couple of notches," and accept the far more reasonable possibility of the existence of a little less powerful Deity? I do not find that in any way necessary to do that myself, because in my worldview – expounded in earlier chapters – the Hitlers and the Stalins of this world do not represent a philosophical enigma. But if they did, I would then "take" the non-omnipotent Creator with finite intelligence *long* before the below-Dumbest-you-know-Who!

And if the anthropomorphic, personal (or individual) God "had to go," why not leave the possibility of a non-anthropomorphic and impersonal Cosmic Intelligence on the table? Einstein, who didn't have "his head screwed on backwards," also rejected the former but accepted the latter – he was very much interested in Spinoza's pantheism – for which he was called an atheist by some of the fine graduates of the Ignoramus Academy (you cannot be a pantheist and an atheist at the same time). I think Einstein's way is an intelligent "way to go." It was this great man of science, who, among others, demonstrated to the world that it is never a good idea to be "too set in one's ways," but to remain flexible in reference to what is so-called "out there." I used the expression "so-called," because there are some recent thinkers who suggest that it may not even be "out there." Be that as it may, what remains certain is that the comfortable pre-1905 worldview of the "well-informed and learned establishment," which told us that we live in an Euclidean universe governed by Newtonian physics, had to give way to the astonishing insight of the Swiss patent office clerk. Einstein's theory of relativity was proven with the observation of eclipses in 1919 and again in 1922. From that theory comes the modern notion that Matter and Energy are not two different realities in essence, and that either one can be converted into the other under certain conditions. Matter is essentially energy congealed into form, which also suggests that higher "forms" of energy can (potentially) re-create the "forms" of matter. That is good news to the one who understands the ramifications of that concept. We all know that cast-iron or cast-steel first has to be melted at a very high temperature before it can assume the shape of the mold of, for example, the car part that we're trying to build. Could it be that Love and joyful Expectancy – being "higher energy" states of consciousness – have a similar capacity to re-create or re-cast the "lower energy" forms, such as Man's conditions of health and prosperity for example, because that might be a legitimate extension of Einstein's discovery in physics. Neat stuff!

Another more recent scientific discovery that I learned about from the TV program called "Nova," suggests that tiny amounts of protons – the positively charged particle in the atom – are routinely "dying" or disappearing out of the Space-Time-Matter (or manifested) Universe. That means that the once firmly held view (which was consistent with the laws of thermodynamics) that the total amount of "Substance (matter, energy in all of its different forms, etc.)" in the Universe remains in a perfect state of equilibrium, or, in other words, all the "Stuff created at the time of the Big Bang is still present in one form or another" – may be out the window too. Wow, stuff disappearing out of the three-dimensional Universe! Is it, after a 13.7 billion year absence, falling back into the possible "supersymmetrical (or parallel) Universes"? What's next? Are all the revered scientific "absolutes" being pulled down like the Berlin Wall in 1989, or the statues of Saddam Hussein in 2003?

I'm not trying to teach a class in physics here (I have no qualifications to do so; I'm only an interested student) – the only reason that I bring this up is that the foregoing clearly tells us that all of this physical stuff that we are so familiar with (or think we are familiar with), and so heavily rely on, may not at all be what we once thought it was.

A familiar question arises again: "Is it all relative, and is there nothing at all that can be classified as absolute? I'm very much aware of the fact that the "relative versus absolute" theme has already been discussed in chapter eight. However, at that point, that theme related to a "value" or a "virtue," where now that same theme, or that same question, pertains to Ultimate Reality, or Life Itself, or the

Rudy Leyerzapf

Truth of the Universe (it's often necessary to use a series of words, since one (single) expression doesn't always adequately describe the reality in question).

Well, since it is beginning to look like the case for (an) *a priori* Intelligence is rock solid, may we now not add that the Grand Motive of this Cosmic Intelligence must be to continue to drive the whole evolutionary process – with all its various stages – forward towards a "higher" life of Beauty, Love and (the right kind of) Power everywhere – but also for, or better through the individual human being playing a tune in the higher octave in particular — and that THAT is absolute? That is one of the few absolute concepts that I will no longer debate in my own mind, where almost everything else is still "up for grabs" with me.

It pleases me that the greatest scientific "discovery" in modern times (Einstein's theory of relativity) has the word "relativity" in it – reminding us to cease to be an absolutist about many things in life.

The evidence of personal transformations

Lord Russell also uses the words: " ... millions of years ...," and " ... final emergence ...." But if we consider the (already used earlier) example of the "conversion" of the – using a modern expression – "low-life" slave trader, who became a "saved" abolitionist, and presumably soon thereafter wrote the well-known song/hymn "Amazing Grace," it should become obvious to any well-thinking individual that we can no longer recognize "the old man" in "the transformed man," and that this transformation may well have taken place "in the twinkling of an eye" rather than in the philosopher's "millions of years." So, the Cosmic Intelligence – from which the possibility of total transformation must have proceeded – may not be anywhere near as handicapped as Bertrand Russell suggested. The slave trader was *not* "the final emergence," but metamorphosed into the far more enlightened abolitionist; and even that version of the individual is not final because of the endless possibilities for further growth.

208

If such transformations are not possible, then we have no choice but to agree with Lord Russell's assessment – but they clearly are possible, and they are happening too, and not at all infrequently. A little too close to the tree, Lord Russell?

Although it is right and legitimate to reflect on these realities momentarily, I have decided long time ago that it is a fine waste of time to wonder too much about Adolf and Joseph, because I have my own life to live instead of a Cosmos to run. The One that's doing that, is doing just fine, and doesn't require my "brilliant assistance."

The rejecting-the-possibility-of-immortality books

Are the bleak picture, randomness-cheerleading authors – I am not including writing about evolution in this question – doing society much, if any, of a favor by putting their rejecting-the-possibility-of-immortality books out in front of the general public? Somebody might say: "That is an arrogant question …, don't they have the same rights as you do?" Well, it is not a question of "having the right," because that exists without argument. The question is: "Are they making a necessary and wholesome – solicited or not – contribution to the reader?" Stephen King makes a very comfortable living "scaring the be-je-bees" – I have never read any of his books and neither do I care to – out of his searching-for-a-rotten-bone readers, but in the end they are supposed to remember that it is fiction. But non-fiction, devoid-of-any-hope bleakness is not inconsequential make-belief, because the possible discomfort derived from it may well be more lasting without that oh-well-it's-just-a-story sigh of relief. In any event, it is much kinder to stay away from that kind of "good feeling" robbery, because as someone once said: "Nobody really knows enough to be a pessimist." Does anyone really think that our current "impressive" human knowledge (taking the liberty of reinforcing what I've said earlier) amounts to anymore than one tenth of one percent – if even that – of what is not-yet-known but eventually knowable; and that may be infinitesimally small compared to what will forever be unknowable. That is not false humility but rather the proper and necessary kind.

When my uncle Joe, who lived thousands of miles away in South Africa, was murdered in the nineteen-fifties, that news was wisely withheld from my already terminally ill grandmother and also from my grandfather who died a few months after her. There was no need to make the gift of horror before their respective transitions. Aren't the treasures of the mind, such as emotional harmony and inner peace, the real much-sought-after and worthwhile "things" in life? Why be the one who shouts: "Hurrah, we've made a great discovery and would "lovingly" like to share it with you: 'there is no future life.'" Is blight equal to its opposite?! No sane person can answer that in the affirmative. If you have nothing hopeful to share, why share it? You could have chosen the much kinder option of keeping it all to yourself! "Yeah, but all the money I could have made …."

Yes, there are times when it is legitimate or even necessary to warn others about certain matters, but the "bleak picture warning" is *not* one of them, because if it all leads to nothingness at the end of life then there is also no following disappointment; or is there?! But the hope before that point *is* real as well as tangible, because as we now know, "positive emotions" and their "happy biochemical followers" can deliver measurable health benefits!

Didn't Carl Sagan (see chapter 2) say that he would prefer to believe that his parents were still alive in another dimension, thereby clearly indicating that the disbelieving or doubting is a less desirable state of mind than its opposite. However, if one must conclude that there isn't enough evidence to embrace the latter (by no means unreasonable), then it still makes far more sense to make the intelligent *and disciplined* "agnostic decision" to suspend judgment on the subject of immortality rather than to adopt the active doubting mode. Why do we always think that we only have two options: believing or disbelieving, when in actuality we clearly have three: believing – suspending judgment – and disbelieving. When they asked Siddhartha Gautama, known as the Buddha, if he believed in immortality, he said he didn't know and indicated that he had banished the speculation from his mind. He neither believed it nor disbelieved it. His – to me superb – advice always was: to learn to live well in the present moment.

When our ten-year-old second daughter died in the nineteen-seventies as a result of meningitis, my wife and I both felt the pressing need to read all the available non-doctrinal literature about the continuity of life that we could get our hands on. When your world seems to have come to an end and your emotions are lying in the gutter, you have no choice but to try to find a way to restore yourself to sanity. I also realized that nothing short of convincing evidence was going "to do much" for me. That was the time when I discovered the great work of Dr. Ian Stevenson (see chapter 5).

Dr. Stevenson's work along with Paramahansa Yogananda's samasamadhi convinced me that we are so much more than flesh and blood, including that wonderful "gray mass" organ we use to figure out the next move on the chess board of life. The veil has now been lifted to some degree. It has been lifted high enough to give Man solid reasons for optimism. It has also been lifted high enough to console the grieving heart. But it also sheer Wisdom and superhuman Brilliance that it hasn't been lifted any higher at this stage, so that He (generic Man) doesn't – what He otherwise would readily have done – become preoccupied with anything other than his present life and his present moments. Apparently our business is here and not some place else.

A great man of insight with a giant intellect, Johann Wolfgang von Goethe – who like Einstein and Spinoza, among others, was also a pantheist – said: "Nothing should be more highly prized than the value of each day." At long last I have decided to make that my motto.

Intelligence, Randomness, and the Stalemate. I give credit to the first one in that line-up to get me past the third one; and the one in the middle only reigns supreme as Lady Luck in the casinos of the world and not in many places outside of them.

About the Author

Rudy Leyerzapf is a graduate of the (now defunct) Washington State chartered University of Metaphysics, a former teacher and seminar lecturer – connected with the Science of Mind Institute, Los Angeles, California, a retired retail store owner, a certified electronics and computer technician, a one-time Chess tournament champion, and a lifelong student of physics, astronomy, history, psychology, and philosophy. He makes his home with his wife of over four decades in Spokane, Washington, and has three children and two grandchildren.

www.ingramcontent.com/pod-product-compliance
Lightning Source LLC
Chambersburg PA
CBHW031945170526
45157CB00002B/396